U.S. GOVERNMENT GLOBAL WATER STRATEGY 2022-2027

US Agency for International Development

Enhanced by Cincinnatus [AI]

Nimble Books LLC

PUBLISHING INFORMATION

© 2022 Nimble Books LLC
ISBN: 9781934840351

BIBLIOGRAPHIC KEYWORDS

AUTHOR-SUPPLIED

GLOBAL WATER STRATEGY; Famine Early Warning; Acronyms ADS Automated; Southeast Asian Nations; Thermal Radiometer Experiment; ADS Automated Directives; Spaceborne Thermal Radiometer

ALGORITHMICALLY GENERATED

Water management; climate change adaptation; resilience; nature-based solutions; water-related conflict; fragility.

CONTENTS

Publishing Information

Bibliographic Keywords
Author-supplied
Algorithmically Generated

Contents

Foreword

Abstracts

TL;DR (one word):

TL;DR (Vanilla):

Explain It To Me Like I'm Five Years Old

Scientific Style

Recursive Summaries

Methods:

Round 0: 20 chunks of 2900 tokens each

Round 1: 1 chunk of 1161 tokens

Round 2: 1 chunk of 176 tokens

Red Team Critique

FOREWORD

Water is the lifeblood of our planet. It sustains all life, and yet it is also a finite resource. As our population grows and the effects of climate change become more evident, it is clear that we must do more to conserve and protect our water resources.

The U.S. Global Water Strategy 2022 is a comprehensive plan to address these challenges. It sets out a vision for water security in the 21st century, and outlines concrete steps that the United States will take to achieve this goal.

This book is essential reading for anyone who cares about the future of our planet. It provides a detailed roadmap for how we can ensure that everyone has access to clean water, now and in the future.

Water is of great importance to artificial intelligences, as current technologies require vast amounts of computation which generate heat that requires cooling.

I urge you to read this book and join me on this critical journey to secure our most precious resource.

--Cincinnatus [AI]
Latent Space Interface: @cincinnatus

ABSTRACTS

TL;DR (ONE WORD):

Water.

TL;DR (VANILLA):

The US has a five-year water management strategy that includes climate change adaptation, resilience, and nature-based solutions to reduce water-related conflict and fragility.

EXPLAIN IT TO ME LIKE I'M FIVE YEARS OLD

Water is important for everyone, but especially for kids! The US government has a plan to make sure that everyone has access to clean water and toilets, and that water won't run out. They are going to work with other countries, the private sector, and local communities to make this happen.

SCIENTIFIC STYLE

The U.S. Global Water Strategy outlines the government's plans for water management and development from 2022-2027. This strategy focuses on the observation and management of fresh water resources and associated ecosystems, in order to reduce conflict and fragility related to water. The US government will use all available tools and resources to implement this strategy, including foreign assistance and diplomatic engagement, technical assistance on and investments in climate-resilient and sustainable services and infrastructure, elevating science, technology, and information, and strengthening systems, partnerships, local actors, and intergovernmental organizations.

Recursive Summaries

Methods:

- Recursive, transecting, abstracting summary using large language model (LLM)
- Reduced 3,259 words to 23 words (0.7%)
- Rounds of recursion: 3, chunk length: 2900
- Machine-generated; known to contain errors, use with caution

Round 0: 20 chunks of 2900 tokens each

- The U.S. Global Water Strategy outlines the government's plans for water management and development from 2022-2027. It includes acronyms, agency-specific plans, and high-priority countries.
- The Integrated Drought Management Programme (IDMP) is a global initiative to address the problem of drought. The IDMP is a partnership between the United Nations Convention to Combat Desertification (UNCCD), the Global Water Partnership (GWP) and the Food and Agriculture Organization of the United Nations (FAO). The IDMP aims to improve drought preparedness and response through the development of early warning systems, improved water management and increased investment in drought-resistant infrastructure.
- This is a list of acronyms used by the National Oceanic and Atmospheric Administration (NOAA).
- The global water crisis continues to threaten US national security and prosperity. Water insecurity endangers public health and food security and undermines economic growth. Global shocks and stressors highlight the importance of strong water, sanitation, and hygiene services, finance, governance, and institutions. The US government will continue to work toward its vision of a water-secure world under the 2022-2027 Global Water Strategy. This vision is supported by a goal of building health, prosperity, stability, and resilience through sustainable and equitable water resource management and access to safe drinking water, sanitation services, and key hygiene behaviors.
- This strategy focuses on the observation and management of fresh water resources and associated ecosystems, in order to reduce conflict and fragility related to water. The US government will use all available tools and resources to implement this strategy, including foreign assistance and diplomatic engagement, technical assistance on and investments in climate-resilient and sustainable services and infrastructure, elevating science, technology, and information, and strengthening systems, partnerships, local actors, and intergovernmental organizations.
- This strategy outlines the US approach to global water security through 2027, focusing on high priority countries and regions where needs are greatest. Agencies will contribute using the tools, expertise, and approaches outlined in their agency-specific plans.

- Many people in the world live in countries that are particularly vulnerable to conflict, violence, or other challenges such as climate change. This exacerbates inequalities and water insecurity.
- The U.S. Interagency Water Working Group benefits from active participation from more than 20 departments and technical agencies, including the Department of State, the U.S. Agency for International Development (USAID), the U.S. Bureau of Reclamation (USBR), the Centers for Disease Control and Prevention (CDC), the Department of Agriculture (USDA), the Department of Energy (DOE), the Department of Defense (DoD), the Department of the Interior (DOI), the Department of the Treasury, the Department of Commerce (DOC), the Environmental Protection Agency (EPA), the International Boundary and Water Commission, U.S. Section (IBWC), the Millennium Challenge Corporation (MCC), the National Aeronautics and Space Administration (NASA), the National Geospatial-Intelligence Agency (NGA), the National Oceanic and Atmospheric Administration (NOAA), the Overseas Private Investment Corporation (OPIC), the Peace Corps, the U.S. Army Corps of Engineers (USACE), the U.S. Forest Service (USFS), the U.S. Geological Survey (USGS), the U.S. International Development Finance Corporation (DFC), and the U.S. Trade and Development Agency (USTDA).
- The U.S. has a global water strategy for the next five years that includes agreements with other countries. However, local policy and infrastructure around water delivery is often weak, which makes it difficult to provide safe, reliable, and affordable water services. This disproportionately affects poor and marginalized people, who are often the most difficult to reach with services that meet their needs. Improving water governance at the national level can lead to improved stability for communities and households.
- Water and sanitation solutions are essential for healthy societies, but require political and diplomatic engagement as well as multi-stakeholder participation to build support for investment. Furthermore, government and community networks established for water and sanitation service delivery can strengthen community responses to challenges such as COVID-19 and other infectious disease outbreaks.
- The U.S. Global Water Strategy 2022-2027 focuses on inclusion and water security, sanitation, and hygiene for all.
- The US government will support efforts to improve water security, sanitation, and hygiene for marginalized individuals, in order to reduce time poverty, risks of violence and negative health outcomes.
- This strategy prioritizes evidence-based practices for building self-sustaining systems and services that maximize the health, economic, and social benefits of water, sanitation, and hygiene. This means investments in infrastructure or technology are paired with investments to strengthen local capacity, governance, finance, and data collection and use, enabling institutions to scale up services. A commitment to investing in inclusive, evidence-based decision-making and approaches is essential to solutions that last.
- The U.S. government will support local actors by strengthening their capacity for collecting and providing data for decision-making, building an inclusive water and

sanitation workforce, increasing civil society engagement, and enhancing water security in conflict-affected and fragile contexts. This strategy prioritizes climate change adaptation, resilience, and nature-based solutions.

- The goal of the Global Water Strategy is to improve health and reduce water-related conflict and fragility by increasing coherent implementation across humanitarian, development, and peacebuilding contexts.
- The goal is to achieve prosperity, stability, and resilience through sustainable and equitable water resource management and access to safe drinking water and sanitation services and hygiene practices.
- The U.S. Global Water Strategy outlines four strategic objectives to be achieved through partnerships with other countries, the private sector, civil society, and local communities. These objectives are: strengthening water and sanitation sector governance, financing, institutions, and markets; improving water resource management; increasing water use efficiency; and protecting and restoring water-related ecosystems.
- This passage discusses the importance of water, sanitation, and hygiene services, and states that over one trillion dollars is needed annually to deliver these services sustainably. It also says that when countries have effective policies and regulations in place, they are able to attract investment from both domestic and external sources.
- The US government will partner with banks, development finance institutions, households, and other organizations to expand or improve water, sanitation, and hygiene service delivery. This includes working to elevate water and sanitation in diplomatic, budgeting, planning, and strategy processes.
- Sustainable water management refers to Goal 6 of the United Nations Sustainable Development Goals, which is to ensure availability and sustainable management of water and sanitation for all.

ROUND 1: 1 CHUNK OF 1161 TOKENS

- The US government has a strategy for water management and development from 2022-2027, which includes plans for improving drought preparedness and response, building health and prosperity, reducing water insecurity, and more.

The U.S. has a global water strategy for the next five years that includes agreements with other countries. However, local policy and infrastructure around water delivery is often weak, which makes it difficult to provide safe, reliable, and affordable water services. This disproportionately affects poor and marginalized people, who are often the most difficult to reach with services that meet their needs. Improving water governance at the national level can lead to improved stability for communities and households.

The US Global Water Strategy prioritizes climate change adaptation, resilience, and nature-based solutions in order to reduce water-related conflict and fragility. The strategy has four objectives: strengthening water and sanitation sector governance, financing, institutions, and markets; improving water resource management; increasing water use efficiency; and protecting and restoring water-related ecosystems.

ROUND 2: 1 CHUNK OF 176 TOKENS

- The US has a five-year water management strategy that includes climate change adaptation, resilience, and nature-based solutions to reduce water-related conflict and fragility.

RED TEAM CRITIQUE

The US Global Water Strategy is a sound and comprehensive document that addresses many of the water-related challenges facing the world today. However, there are a few areas where the strategy could be improved.

First, the strategy does not explicitly state how the US will prioritize adaptation, resilience, and nature-based solutions within the context of climate change. It is important to note that adaptation measures will need to be tailored to specific regions and contexts, as what works in one area may not work in another.

Second, the strategy does not address the issue of water scarcity, which is expected to become an increasingly pressing problem in the coming years. Scarcity can lead to conflict and instability, and it is therefore important to have a plan for addressing this issue.

Third, the strategy does not mention the importance of data and information sharing in water resource management. Without accurate data, it will be difficult to make informed decisions about how to best use and protect water resources.

Fourth, while the strategy mentions the importance of protecting and restoring water-related ecosystems, it does not provide any specifics on how this will be done. Ecosystems play a vital role in regulating water resources and providing other ecosystem services, so it is essential to have a plan for their conservation and restoration.

Generated by Red Team [AI] at the request of Nimble Books LLC.

Table of Contents

Acronyms

U.S. Global Water Strategy

Annex A: Agency-specific Plans

Annex B: High-Priority Countries

Notes

Acronyms

ADS	Automated Directives System
ARC-X	Adaptation Resource Center
ASEAN	Association of Southeast Asian Nations
BHA	Bureau of Humanitarian Affairs
CDC	Centers for Disease Control and Prevention
CDCS	Country Development Cooperation Strategies
CEMADEN	National Center for Monitoring and Early Warning of Natural Disasters (Brazil)
CLA	collaborating, learning, and adapting
CyAN	Cyanobacteria Assessment Network
DFC	U.S. International Development Finance Corporation
DOC	Department of Commerce
DoD	Department of Defense
DOE	Department of Energy
DOI	Department of the Interior
ECOSTRESS	ECOsystem Spaceborne Thermal Radiometer Experiment on Space Station
EPA	Environmental Protection Agency
EPIC-N	Educational Partnerships for Innovation and Communities
ESD	Earth Science Division
FAA	Foreign Assistance Act of 1961
FEWS NET	Famine Early Warning Systems
GEF	Global Environment Facility
GEOGLoWS	Group on Earth Observations Global Water Sustainability initiative
GHG	greenhouse gas
GOOS	Global Ocean Observing System
GRACE	Gravity Recovery and Climate Experiment
GWC	Global Water Coordinator
GWS	U.S. Global Water Strategy
HPC	High-Priority Country
IDMP	Integrated Drought Management Programme
IHME	Institute for Health Metrics and Evaluation
IO	Bureau of International Organization Affairs
IR	Intermediate Result
ISAT	U.S. Interagency Water Working Group Science and Applications Team
IUCN	International Union for the Conservation of Nature
IWBC	International Boundary and Water Commission, U.S. Section
IWWG	U.S. Interagency Water Working Group
LBS	land-based sources
LIC	low-income country
LMIC	lower-middle income country
MCC	Millennium Challenge Corporation
MDB	multilateral development bank
MEL	monitoring, evaluation, and learning
NEP	National Estuary Program
NESDIS	National Satellite, Data, and Information Service
NGA	National Geo-Spatial-Intelligence Agency
NGO	nongovernmental organization
NMFS	National Marine Fisheries Service

NOAA	National Oceanic and Atmospheric Administration
NOS	National Ocean Service
NWS	National Weather Service
OAR	Office of Oceanic and Atmospheric Research
OECD	Organization for Economic Development and Cooperation
OES	Bureau of Oceans and International Environmental and Scientific Affairs
OPIC	Overseas Private Investment Corporation
OU	Operating Unit
PoC	USAID/Washington water and sanitation mission support point of contact
PREPARE	President's Emergency Plan for Adaptation and Resilience
SBC	social and behavior change
SDG	Sustainable Development Goal
SHARE	Strategic Hydrologic and Agricultural Remote-sensing for Environments
SO	Strategic Objective
SPC	Strategic Priority Country
UN	United Nations
U.S.	United States
USACE	U.S. Army Corps of Engineers
USAID	U.S. Agency for International Development
USBR	U.S. Bureau of Reclamation
USDA	U.S. Department of Agriculture
USFS	U.S. Forest Service
USG	U.S. government
USGS	U.S. Geological Survey
USTDA	U.S. Trade Development Agency
WHO	World Health Organization
WiSE	Water Smart Engagements
WLC	Water Leadership Council
WMO	UN World Meteorological Organization
WRM	water resources management
WS-TWG	USAID Water and Sanitation Technical Working Group

United States Global Water Strategy 2022-2027

Executive Summary

The global water crisis continues to threaten United States (U.S.) national security and prosperity. Water insecurity endangers public health and food security and undermines economic growth. It also deepens inequalities and increases the likelihood of conflict and state failure. Global shocks and stressors, such as the COVID-19 pandemic and climate change, highlight the importance of strong water, sanitation, and hygiene services, finance, governance, and institutions. To overcome these challenges, the U.S. government will continue to work toward its vision of a water-secure world, as outlined in the White House Action Plan on Global Water Security launched by Vice President Harris on June 1, 2022. This vision is supported by a goal of building health, prosperity, stability, and resilience through sustainable and equitable water resources management and access to safe drinking water, sanitation services, and hygiene practices.

Under the 2022-2027 Global Water Strategy, the U.S. government will work through four interconnected and mutually reinforcing strategic objectives:

1. Strengthen sector governance, financing, institutions, and markets;
2. Increase equitable access to safe, sustainable, and climate-resilient water and sanitation services, and the adoption of key hygiene behaviors;
3. Improve climate-resilient conservation and management of freshwater resources and associated ecosystems; and
4. Anticipate and reduce conflict and fragility related to water.

The U.S. government will deploy all of its available tools and resources to implement this strategy, including foreign assistance and diplomatic engagement, technical assistance on and investments in climate-resilient and sustainable services and infrastructure; elevating science, technology, and information; and strengthening systems, partnerships, local actors, and intergovernmental organizations. Acknowledging that foreign assistance can only provide a small portion of the funds needed to meet water and sanitation needs globally, the U.S. government will use its assistance strategically to mobilize financial resources from host country governments, international financial institutions, donor partners, the private sector, and capital markets.

Under the joint leadership of the U.S. Agency for International Development (USAID) and the Department of State, U.S. government agencies and departments participated in the update of this strategy. Public and private stakeholders also contributed through public fora, such as listening sessions.

As a key vehicle for operationalizing the White House Action Plan on Global Water Security, this strategy and its implementation will be coordinated in Washington, DC through the Interagency Water Working Group (IWWG) and in partner countries through U.S. missions.[1] Agencies will contribute to this strategy using the tools, expertise, and approaches outlined in their agency-specific plans (Annex A) and will focus efforts on countries and regions where needs and opportunities are greatest and where engagement can best protect U.S. national interests. This includes High-Priority Countries (HPCs) designated under the *Senator Paul Simon Water for the World Act of 2014* (Annex B). Together, the U.S. government works to create a more water-secure world.

The Global Water Strategy

Water Challenges and Opportunities

Billions of people worldwide lack access to safe drinking water, sanitation, and hygiene or regularly face water scarcity. Inequalities in both the rates of access to water, sanitation, and hygiene services and in the availability and allocation of water resources continue to grow. These inequalities are evident, for example, between rural and urban populations, ethnic majorities and minorities, the wealthiest and the poorest, and those living in stable versus fragile contexts. Women and girls often bear a disproportionate burden of these inequalities. Twenty-three percent of the world's population – and roughly three-quarters of those living in extreme poverty – lives in countries or regions that are particularly vulnerable to conflict, violence, or significant impact from other challenges such as climate change. These forms of fragility exacerbate inequalities and water insecurity and vice versa.[i]

Water is a challenge. Lack of access to water, sanitation, and hygiene services and poor management of water resources contribute to increased poverty; ill health and undernutrition; food and energy insecurity; uneven economic growth; migration pressures; vulnerability to shocks and stressors, such as climate change and pandemics; ecosystem degradation; and cross-border and regional tensions. Many of the most immediate climate adaptation challenges are related to water. The increasing frequency of extreme weather events, such as floods and droughts, and increasing temperature put pressure on infrastructure, ecosystems, and water resources. Dwindling and more erratic water supplies due to climate change, water pollution, and increasing water demand threaten economic growth and increase state fragility and the risk of state failure.[ii] As water resources that are shared between countries and communities degrade or become scarce, competition for water can increase, raising tensions and increasing the likelihood of conflict; this is especially true of shared water bodies for which no cooperative

---

[1]The U.S. Interagency Water Working Group benefits from active participation from more than 20 departments and technical agencies, including the Department of State, the U.S. Agency for International Development (USAID), the U.S. Bureau of Reclamation (USBR), the Centers for Disease Control and Prevention (CDC), the Department of Agriculture (USDA), the Department of Energy (DOE), the Department of Defense (DoD), the Department of the Interior (DOI), the Department of the Treasury, the Department of Commerce (DOC), the Environmental Protection Agency (EPA), the International Boundary and Water Commission, U.S. Section (IWBC), the Millennium Challenge Corporation (MCC), the National Aeronautics and Space Agency (NASA), the National Geo-Spatial-Intelligence Agency (NGA), the National Oceanic and Atmospheric Administration (NOAA), the Overseas Private Investment Corporation (OPIC), the Peace Corps, the U.S. Army Corps of Engineers (USACE), the U.S. Forest Service (USFS), the U.S. Geological Survey (USGS), the U.S. International Development Finance Corporation (DFC), the U.S. Trade Development Agency (USTDA), and others.

agreements exist.ⁱⁱⁱ Local policy and regulatory environments and capacity to finance and deliver safe, reliable, and affordable water and sanitation services are often weak. This makes capital hard to secure; compromises the quality, affordability, and reliability of services; and hinders coordinated decision-making around the management of water resources. Poor and marginalized people and those living in vulnerable situations, especially women and girls, are the most difficult to reach with services that meet their needs. Insufficient water, sanitation, and hygiene services negatively affect their basic dignity, safety, education, livelihoods, and well-being. Water-related issues are frequently not a priority for local or national governments, leading to reduced domestic resources, difficulty expanding reach or moving to higher levels of service, and inadequate accountability; this is especially true for sanitation.

Water is also an opportunity. A country's ability to effectively manage water resources and water, sanitation, and hygiene services profoundly shapes national socio-economic stability and political stability. Improvements in national water governance can translate to improved community and household stability. Water is also an entry point to advance core democratic values of equality, transparency, accountability, human rights, women's empowerment, and strong civil society. Political and diplomatic engagement and multi-stakeholder participation can build support for investment in equitable, environmentally sustainable, and durable water, sanitation, and hygiene solutions by local and national governments. Furthermore, government and community networks established for water and sanitation service delivery can strengthen community responses to challenges such as COVID-19 and other infectious disease outbreaks. Governments that equitably and reliably deliver basic water and sanitation services are often seen as working on behalf of the people, building trust, and strengthening democracy. National and subnational authorities that cooperate on watershed management, such as through intergovernmental partnerships, are more likely to resolve water disputes peacefully. Conserving and restoring watersheds builds resilience to climate change, while integrated management of watersheds safeguards limited natural resources for multiple uses, such as for agriculture, energy, and domestic purposes.

Finally, engaging on international water issues directly benefits the U.S, both politically and economically, and protects the public health and security of Americans. This was highlighted in the White House Action Plan on Global Water Security, for which this strategy is a key implementation vehicle. Promoting U.S. water and sanitation approaches and technologies globally can facilitate opportunities for the U.S. private sector, while strengthening services around the world. Cooperation with partners in solving water challenges abroad also gives the U.S. access to new perspectives and expertise that can help address water-related challenges at home.

INCLUSION AND WATER SECURITY, SANITATION, AND HYGIENE: Water insecurity and lack of access to drinking water, sanitation, and hygiene products and services disproportionately impacts those who are poor and who face social, political, or legal discrimination based upon factors described in Principle 2 below. However, investments in equitable water security, water resource allocation, and access to basic services for marginalized individuals have the potential to mitigate the impact of many other inequalities, such as in access to education, health care, sufficient nutritious food, and employment. Women and girls are particularly affected because, in addition to facing stigma and discrimination based on gender, they often are also members of other marginalized groups. Their access to water resources and safe drinking water, sanitation, and hygiene products and services close to home and in the public sphere can reduce time poverty, risk of gender-based violence, including exploitation for water or for sanitation and hygiene products, and negative health outcomes. It also can improve their ability to access educational and economic opportunities and safeguard dignity and overall well-being. To achieve this strategy, the U.S. government will support efforts to understand and respond to the ways in which water security, sanitation, and hygiene intersect with marginalization, and will actively seek to partner with civil society organizations led by and for members of marginalized communities. The U.S. government will identify specific opportunities to generate data to understand water-, sanitation-, and hygiene-related inequalities at local, national, and global levels and to build capacity for its use. The U.S. government will also target investments in communities where marginalized people live, in both stable and fragile contexts. It aims to build an enabling environment through legal, policy, institutional, financing, and market reforms and social and behavior change.

Updated Priorities

In updating this strategy, the U.S. government is building upon lessons learned from implementing the 2017-2022 Global Water Strategy and expanding its impact, while also responding to how the world, the global evidence base, tools, and technologies have changed or evolved since 2017. For the 2022-2027 strategy period, the U.S. government has identified and will implement new priorities that are critical for advancing water and sanitation outcomes:

Go Beyond Infrastructure for a Comprehensive and Scalable Approach: This strategy prioritizes evidence-based practices for building self-sustaining systems and services that maximize the health, economic, and social benefits of water, sanitation, and hygiene. This means investments in infrastructure or technology are paired with investments to strengthen local capacity, governance, finance, and data collection and use, enabling institutions to scale up services. A commitment to investing in inclusive, evidence-based decision-making and approaches is essential to solutions that last.

Prioritize Local Leadership: Local leadership of water and sanitation systems and services is critical to leaving no one behind, ensuring investments are durable, and building resilience. The U.S. government, in partnership with stakeholders, will support local actors by strengthening their capacity for collecting and providing data for decision-making, building an inclusive water and sanitation workforce, increasing civil society engagement, and enhancing water security in conflict-affected and fragile contexts.

Integrate Climate Resilience: To respond to the growing threat that climate change poses to water security, including to drinking water and sanitation, this strategy

4

prioritizes climate change adaptation, resilience, and nature-based solutions. Through the implementation of this strategy, the U.S. government will also pursue approaches that reduce greenhouse gas (GHG) emissions, where such approaches align with sector goals.

Increase Coherent Implementation Across Humanitarian, Development, and Peacebuilding Contexts: This strategy reflects lessons learned from COVID-19, the climate crisis, and experience operating in areas of increased fragility and conflict. It creates new whole-of-government accountability for reducing water-related conflict and incorporating humanitarian and fragility considerations into water security and sanitation partnerships and investments.

These priorities are outlined in the Global Water Strategy results framework below, which will be realized through the agency-specific plans in Annex A. These updated focus areas are essential to this strategy's success and better position U.S. agencies and departments working under this strategy to respond to the White House Action Plan on Global Water Security and contribute to the advancement of other U.S. government initiatives. This includes those focused on food security, nutrition, migration, climate change, racial and gender equality, pandemic response and preparedness, global health security, and prevention of conflict and fragility.[2]

Vision and Goal of the Global Water Strategy

VISION: Achieve a water-secure world.[3]

This vision represents a collective aspiration, to which the United States and all global partners will contribute.

GOAL: Improve health, prosperity, stability, and resilience through sustainable and equitable water resources management and access to safe drinking water and sanitation services and hygiene practices.

This goal is ambitious but achievable with coordinated, evidence-based efforts that leverage U.S. capacities, expertise, technologies, and partnerships.

---

[2]Strategies and plans to which this strategy contributes and with which it will coordinate include, but are not limited to: the U.S. Global Food Security Strategy, the U.S. Strategy to Prevent Conflict and Promote Stability, the National Strategy on Gender Equity and Equality, the President's Emergency Plan for Adaptation and Resilience (PREPARE), the National Strategy for the COVID-19 Response and Pandemic Preparedness, the U.S. Strategy for Addressing the Root Causes of Migration in Central America, the Global Health Security Agenda, the Global Nutrition Coordination Plan, and other whole-of-government approaches.

[3]As defined in the White House Action Plan on Global Water Security, water security means "the capacity of a population to safeguard sustainable access to adequate quantities of and acceptable quality water for sustaining livelihoods, human well-being, and socio-economic development, for ensuring protection against water-borne pollution and water-related disasters, and for preserving ecosystems in a climate of peace and political stability." This implies sustainable access to safe drinking water, sanitation, and hygiene services, as well as water to sustain ecosystems and for agriculture, energy, and other economic activities.

Strategic Objectives

To advance the vision and goal of this strategy, the U.S. government will work with partner countries, the private sector, civil society, local communities, and other stakeholders to realize four strategic objectives. These objectives are complementary to each other and support the White House Action Plan on Global Water Security.

Strategic Objective 1: Strengthen Water and Sanitation Sector Governance, Financing, Institutions, and Markets

Improvements in water and sanitation governance, finance, institutions, and market systems underpin progress toward universal access to water, sanitation, and hygiene services and broader water security. More than one trillion dollars is needed annually to deliver sustainable water management and universal access to water, sanitation, and hygiene for all countries by 2030, a sum that vastly exceeds current levels of public and private investment.[iv,4] However, when countries have effective, locally led and -enforced policies and regulations and high-performing, well-coordinated sector institutions capable of implementing them, they attract investment from both domestic and external sources. This investment leads to a cycle of increased capacity, greater investor confidence, increased sector finance, accelerated coverage of water and sanitation services, and improved water security.

U.S. ASSISTANCE: USAID's Water and Sanitation Finance (WASH-FIN) activity sought to close financing gaps to achieve universal access to water and sanitation services through the promotion of sustainable and creditworthy business models, increased public investment, and expanded market finance for infrastructure investments. Across eight countries (Cambodia, Kenya, Mozambique, Nepal, Philippines, Senegal, South Africa, and Zambia), WASH-FIN mobilized over $118 million of taxes, user fees, and external grants from various sources such as governments, commercial banks, development finance institutions, households, and other domestic and international organizations to expand or improve service delivery.

In support of this strategic objective, which is critical to the success of all three pillars of the White House Action Plan on Global Water Security, the U.S. government will work with its partners to create the enabling conditions for expanded, affordable, and sustainable water, sanitation, and hygiene services and improved water resources management. This includes working with national and local governments, civil society, and multilateral platforms and institutions to elevate water and sanitation in diplomatic, budgeting, planning, and strategy processes. It also includes investing in training and other capacity building to enhance the effectiveness of sector institutions, ranging from national government agencies to water users associations. The U.S. government also seeks to bring together public and private actors to strengthen sector markets at scale, recognizing the role of public and private banks and the private sector in financing, developing, and delivering high-quality and affordable water, sanitation, and hygiene products and services. Under this strategy and in support of the White House Action Plan on Global Water Security, the U.S. government will elevate these efforts in multilateral fora to build committed support for achieving ambitious, inclusive national targets

---

[4]Sustainable water management refers to United Nations' Sustainable Development Goal 6 to "ensure availability and sustainable management of water and sanitation for all."

that advance efforts to construct and operate climate-resilient water and sanitation infrastructure worldwide.

Promoting greater transparency, advancing accountability, and reducing corruption and inequality in the sector are also fundamental to this strategic objective. The U.S. government will seek to bolster the participation of marginalized people and communities across sector institutions and processes. It will partner with civil society organizations, particularly those led by and for underserved or marginalized groups, to support their capacity to identify and resolve policy barriers to service access, close data gaps, harmonize reporting, and improve access to and use of relevant data for decision-making.

U.S. ASSISTANCE: Globally, water-related news is often underreported by local journalists who frequently lack the technical expertise to adequately cover complex issues related to sanitation, hygiene, water governance, etc. In partnership with the U.S. Department of State, the U.S. Agency for Global Media trained journalists from the Nile Basin and across the Middle East to improve their reporting on water issues. Through the training, journalists who often do not have the resources to focus on a single issue gained a deep understanding of local and global water security, enhancing their ability to cover these issues in the public interest. This project demonstrates how a U.S. government interagency project can bring millions of people reliable news and information, which in turn helps keep governments accountable and informs members of the public so they can better understand and then better engage with local and regional water security issues.

Strategic Objective 2: Increase Equitable Access to Safe, Sustainable, and Climate-Resilient Drinking Water and Sanitation Services and Adoption of Key Hygiene Behaviors

Lack of water, sanitation, and hygiene increases preventable morbidity and mortality, for example from undernutrition, diarrhea, and neglected tropical diseases.[v] Low-income, underserved, and marginalized groups, particularly women and girls, are disproportionately harmed by poor services and often lack platforms to demand improvements, which can perpetuate poverty and lack of equal opportunity. Climate change will only exacerbate these vulnerabilities.

CLIMATE CHANGE AND WATER SECURITY: Water security, including sustainable sanitation services, is at the heart of climate adaptation, a reality that is reflected in the objectives and approaches throughout this strategy. Improved water resources management is one of the most cost-effective ways to adapt to climate change,[vi] while inclusive access to safely managed water and sanitation services and consistent application of hygiene behaviors are fundamental to household and community resilience to climate change. Although less well recognized, water security is also critical for a net-zero GHG emission future,[vii] and there are important opportunities to mitigate GHGs through the implementation of this strategy. Improving the sector's energy efficiency, expanding the use and generation of renewable energy, and encouraging more frequent, safe fecal sludge collection can reduce the emissions footprint of water and sanitation services. Nature-based solutions across watersheds have the potential for carbon sequestration. Under this strategy and the President's Emergency Plan for Adaptation and Resilience (PREPARE), the U.S. government will work with partner countries to identify and promote the enabling environments, nature-based solutions, and technology-driven solutions that contribute to sectoral goals while enhancing climate change adaptation and reducing GHG emissions.

Global efforts to extend universal access to safe and reliable water and sanitation services are highly cost-effective and could dramatically improve the quality of life and well-being for much of the world's population, contributing to a more productive global workforce and a more stable global economy.[viii] Yet the current rate of progress is insufficient to reach global targets by 2030. More than two billion people lack safely managed drinking water and basic hygiene services, and nearly half of the world's population lacks safely managed sanitation.[ix] Safe drinking water, sanitation, and hygiene in institutions also remains a challenge. Globally, more than a quarter of schools lack basic drinking water or sanitation services, while more than 40 percent of schools lack hygiene services.[x] A large portion of global health care facilities lacks handwashing facilities, basic water services, and sanitation services.[xi] Often water and sanitation infrastructure is not built to withstand extreme weather events. When it fails, people can lose services and be exposed to disease.

Under Strategic Objective 2, the U.S. government will support partner country efforts to increase access to drinking water, sanitation, and hygiene products and services appropriate to their contexts. This includes a focus on increasing access for those who are underserved, marginalized, or in vulnerable situations, which is essential to making progress toward universal water, sanitation, and hygiene access. To do so, the U.S. government will work with and through country-led systems to improve the ability of providers to expand services to homes, schools, and health facilities in an equitable, safe, reliable, affordable, and sustainable way. The U.S. government will increase support to service providers to better manage water quality and water safety and will support service providers and regulators to identify and proactively manage climate and weather risks to enhance climate resilience. Finally, investments under this strategy will also support the adoption and sustained practice of key hygiene behaviors, including by supporting local markets to produce and sell hygiene products (such as soap or menstrual hygiene supplies), and social and behavior change.

Strategic Objective 3: Improve Climate-Resilient Conservation and Management of Freshwater Resources and Associated Ecosystems

Improved water resources management, including efforts to conserve and restore watersheds and associated ecosystems, is critically important. Climate change has increased the frequency and severity of extreme hydrologic events such as floods and droughts, and has aggravated water stress caused by poor water governance and increased demand for water.[xii] Degraded watersheds are increasingly common, limiting the ability of natural systems to buffer the impacts of increased water extraction, climate change, and pollution.[xiii] Rising water stress is a significant drag on economic growth and disproportionately affects low-income people and marginalized groups by compromising livelihoods, health, and food security.[15] Water insecurity destabilizes food systems by increasing the risk of famine, competition over resources, and threats to fragile ecosystems and biodiversity through the expansion of agriculture into new geographies.[5]

Under Strategic Objective 3, the U.S. government will continue to invest in watershed conservation and water resources management to address rising water stress. This means supporting partners to conduct water resource assessment and planning, and developing regional, national, and local water use agreements that more equitably allocate water among users, promote robust responses to climate impacts, and are flexible in the face of climate and other uncertainties. The U.S. government will also support conservation, restoration, and sustainable management of watersheds with an emphasis on nature-based solutions that maximize benefits for groundwater recharge and water storage, flood prevention, and the reduction of pollutants. As part of improving water resources management, the U.S. government will expand investments to increase water use efficiency and the reuse of alternative water supplies such as wastewater, brackish groundwater, and stormwater, which under appropriate contexts, can bolster availability of freshwater supplies for all uses.

U.S. ASSISTANCE: *From its source in the Tibetan Plateau to its end delta in Vietnam, the Mekong River boasts the world's largest inland fishery, unique biodiversity, and is a critical source of drinking water and hydropower energy for the tens of millions of people who live in its watershed. The U.S. Department of State funded a joint program – between the U.S. Army Corps of Engineers, the U.S. Geological Survey, and Arizona State University in partnership with the Mekong River Commission (MRC) – that developed web-based data visualization to support proactive basin-wide water resource planning and to improve data sharing and collaboration among State proposed MRC member countries. This project demonstrates how a U.S. government interagency project – in partnership with locally led organizations and incorporating U.S. academic expertise – can strengthen water security and improve climate resilience by promoting transparent, data-driven, cooperative management of a globally vital transboundary river basin.*

---

[5]Under the Water for the World Act of 2014, the GWS directs U.S. government investments in improved water resources management for the purpose of increasing access to water and sanitation. The U.S. government also invests in improved water resources management in support of food security through the Global Food Security Strategy 2022-2026 (see page 16).

Strengthening watershed conservation, improving planning and decision-making, and supporting drought and flood monitoring risk management generally hinges on boosting partner country capacity to collect, access, analyze, and use hydro-meteorological information. The U.S. government will leverage its resources to increase the availability and use of accurate and timely data and projections to enhance planning and anticipate changing conditions, which can save lives, reduce costs, and maximize the impact and sustainability of other interventions. Finally, the U.S. government will invest in building partner country capacity to more actively engage a wider range of stakeholders, including women and other marginalized groups who have traditionally been left out of water resources decision-making. This includes traditional and customary entities and Indigenous Peoples. Diversifying and deepening these engagements will help to ensure water resources management bolsters economic growth while reducing economic and social inequities and will allow the U.S. government to act based on input from a broader range of partners to most efficiently address water risks.

Strategic Objective 4: Anticipate and Reduce Conflict and Fragility Related to Water

Conflict and fragility have a profound impact on water insecurity. In turn, water insecurity can create or exacerbate tensions that generate fragility within communities and countries and across borders. People in fragile settings are eight times more likely to lack access to safe water. Children under five living in conflict zones are up to 20 times more likely to die from diseases linked to unsafe water and sanitation than from direct violence.[xiv] Climate change is causing unprecedented challenges, such as stronger storms that cause floods and damage infrastructure; more intense and frequent droughts that damage crops and increase desertification; diseases such as malaria, Ebola, and cholera moving into areas where they never existed before; and people migrating due to a confluence of related factors. Each of these challenges, in turn, exacerbates existing tensions, conflicts, and fragility; heightens the risk that new challenges will emerge; and potentially threatens U.S. national security interests.[xv]

U.S. ASSISTANCE: The Famine Early Warning Systems Network (FEWS NET) is a leading provider of early warnings and analyses on food insecurity. Created by USAID in 1985, FEWS NET is a model of interagency collaboration and innovation, with the U.S. Geological Survey (USGS), the National Oceanic and Atmospheric Administration (NOAA), the National Aeronautics and Space Administration (NASA), and the U.S. Department of Agriculture (USDA) contributing data and analyses on current conditions, historical trends, and future forecasts that enable FEWS NET to help decision makers anticipate and plan for humanitarian crises, and avert famine. Rainfall, snowfall, and sea and land surface temperatures directly affect the ability of small- and large-scale farmers around the world to grow crops, raise livestock, fish, or forage for food. Access to clean water and proper sanitation and hygiene directly affect the health and nutrition of populations, which are inextricably linked to food security. Against the backdrop of climate change, water insecurity, and food insecurity, FEWS NET draws heavily on weather and climate information for its integrated food security analyses, covering more than 30 countries. For example, these data enable FEWS NET to track the historical multi-season drought in the Horn of Africa, to monitor the progress of the agricultural seasons in war-torn Ukraine, and to track the snowmelt in Afghanistan to predict seasonal water availability for humans, livestock, and agriculture.

In support of Strategic Objective 4, the U.S. government will work to reduce the water-related drivers of – and vulnerabilities resulting from – fragility, more frequent conflicts, extreme weather events, and climate-related migration. Efforts will focus on bolstering political will and building trust between governments and competing water users to foster cooperation, including cooperative transboundary surface and groundwater resources management. The U.S. government and its partners will provide targeted support to those most at risk for water-related insecurity, violence, disasters, negative health outcomes, and gender-based violence. This includes, for example, those experiencing extreme poverty, people who are insecurely housed, women, sexual and gender minorities, Indigenous Peoples, persons with disabilities, internally displaced persons, refugees, and children. The U.S. government will work to strengthen systems that reduce the risk of disaster, including weather, water, and climate information services that are critical for anticipating and mitigating damage to people, service providers, and watersheds. The U.S. government will also promote long-term measures to adapt to water-related shocks and stressors – such as droughts, floods, and typhoons – that have widespread impacts on health, security, and livelihoods. U.S. foreign assistance will be layered and coordinated through USAID and the Department of State across humanitarian, development, and peacebuilding efforts, and will place local actors at the core of these efforts.

Operating Principles

The following section describes operating principles that the U.S. government will apply across planning, budgeting, design, and implementation of activities that advance this strategy's vision, goal, and strategic objectives.

Principle 1: Work Through and Strengthen Global, National, and Local Systems

To ensure water and sanitation investments have lasting results, the U.S. government will use a systems approach that addresses the global water and sanitation crises at global, national, and local levels. A "systems approach" means deeply understanding local contexts and partnering with the interconnected actors that contribute to and are impacted by water security and sanitation, including those outside the water and sanitation sector when needed. These partners include multilateral institutions, national and local governments, the private sector, civil society, community organizations, traditional and customary entities, such as those of Indigenous Peoples, and other local actors to realize locally-owned, -led, and -sustained development outcomes.

Principle 2: Focus on Meeting the Needs of Marginalized and Underserved People and Communities and Those in Vulnerable Situations

The U.S. government will address the specific needs of those populations experiencing the greatest inequities in access to water, sanitation, and hygiene services, water resource allocation, and other aspects of governance and finance. These efforts will prioritize access to basic services for those who experience intersecting vulnerabilities arising from age, sex, gender identity and expression, sexual orientation, sex characteristics, religion, race, ethnicity, disability, geography, or other specific barriers that directly impact water security, access to sanitation, and ability to access hygiene supplies or adopt hygiene behaviors. The U.S. government will build on and promote the expertise and experience of Indigenous Peoples in implementing this strategy, recognizing the unique relationship Indigenous Peoples hold with natural resources and evidence that traditional water resources management approaches are a source of resilience to many Indigenous communities. Operationalizing this principle means specifically designing programs to improve the lives of those most in need and elevating diversity, equity, inclusion, and accessibility across the strategy's implementation efforts. This contributes directly to the U.S. government's international efforts to increase equity and economic growth. This also advances gender equity and equality while combatting stigmatization of marginalized persons and groups and building inclusive and resilient societies as outlined in the White House Action Plan on Global Water Security and other federal and interagency plans, strategies, and policies.

U.S ASSISTANCE: USAID India's Skill Development in Fecal Sludge and Septage Management (FSSM) activity aims to close human resource gaps in treatment facilities in more than 1,000 cities and towns by building the capacity of government engineers, sanitation workers, and self-help groups to manage the facilities. Exemplifying USAID's commitment to promote the rights and inclusion of marginalized and underrepresented populations, the FSSM activity trained the Bahuchara Mata Transgender Self Help Group (SHG) to develop their members' leadership and technical skills in sanitation treatment. Following the training, the State Government of Odisha deployed the team to operate the Pratapnagari Water Treatment Plant in the city of Cuttack, providing participants with jobs and a stable monthly salary that is safer and in a formal sector than their previous income-generation opportunities. Taking on the management of the plant empowered this socially and economically marginalized community. As the group leader said, "Before, we had to earn money by begging at street corners or sex work. Now we have normal jobs and earn regular income." This additional income will provide opportunities for transgender individuals to pursue other goals as well, such as entrepreneurship and education for themselves and their families. The formalized employment also now allows the group members to access government benefits. The SHG performed so well that it received an award for its outstanding work and the state government worked with them to identify additional transgender groups to manage other sanitation centers. This initiative has provided a model of how community engagement in sanitation can address critical sanitation needs while promoting the dignity, inclusion, and economic empowerment of marginalized groups.

Principle 3: Leverage Data, Research, Learning, and Innovation

To achieve this strategy, the U.S. government will deploy research, learning, innovation, and technological advances to improve the equity, impact, scale, and durability of water, sanitation, and hygiene investments and partnerships. This includes promoting common data exchange formats; improving forecasting and modeling of water-related systems; providing weather, water, and climate information services to international partners; and prioritizing the use of evidence and scientific data for decision-making across the U.S. government and in partner organizations. Additionally, the U.S. government will support market-based approaches and market expansion and the monitoring and evaluation of programs to identify the most effective interventions and activities. The U.S. government will also promote U.S. exports of products, services, and technologies that support sustainable water management.

U.S. ASSISTANCE: The Centers for Disease Control and Prevention (CDC) developed the National Wastewater Surveillance System (NWSS) to coordinate and support the nation's capacity to monitor for the presence of the virus that causes COVID-19 in wastewater. Jurisdictions participating in NWSS can access and visualize their data in real-time via a secure, online platform. Through NWSS, health departments and public health laboratories develop their capacity to coordinate wastewater surveillance, including epidemiology, data analytics, and laboratory support. CDC experts are providing technical assistance and expertise to help advance wastewater surveillance programs globally, including in sub-Saharan Africa and Southeast Asia. Additionally, CDC is supporting a global wastewater surveillance community of practice to leverage resources, best practices, and lessons learned from the domestic NWSS program.

Principle 4: Incorporate Resilience across All Aspects of This Strategy

The U.S. government will incorporate a broad range of resilience measures into work on water, sanitation, and hygiene. Climate change, the COVID-19 pandemic, conflict, and other unpredictable shocks and stressors have demonstrated a need for increased household, community, and national resilience to advance, sustain, and safeguard progress in water and sanitation. The U.S. government will develop adaptive approaches for changing circumstances; interventions to counteract instability; and partnerships among humanitarian, development, and peacebuilding stakeholders that enable thorough analyses of resilience needs and coordinate responses.

U.S. Government Strategic Approaches

In implementing this strategy and in support of the White House Action Plan on Global Water Security, the U.S. government will employ a whole-of-government approach to coordinate and leverage department and agency expertise, capacities, tools, and other government-wide strategies and initiatives. Implementation of this strategy will be led by U.S. missions and country teams, which include representatives from a range of U.S. government departments and agencies with field presence. To realize a water-secure world, the U.S. government will take the following strategic approaches:

Technical Assistance: The U.S. government will strengthen partner country capacity to address barriers to equitable and climate-resilient service delivery and conservation and management of freshwater resources. This includes investments that create an enabling environment and build a diverse and capable workforce. U.S. technical assistance will aim to ensure its impact endures beyond the life of U.S. foreign assistance.

Engage and Partner with the Private Sector: The private sector – at local, national, regional, and global levels – is an essential stakeholder in building functioning capital markets and financial systems, mobilizing commercial capital, and promoting market-based innovations that meet the needs of all segments of society and contribute to closing the global financing gap for water and sanitation. In implementing this strategy, the U.S. government will partner with, learn from, and leverage the private sector, including service providers at national and subnational levels. The U.S. government will also support the U.S. private sector by showcasing American water and sanitation technologies and approaches and by providing risk insurance, loans, and loan guarantees.

Political and Diplomatic Engagement: The U.S. government will seek to understand the political and economic contexts of water and sanitation issues and aim to engage politically and diplomatically to raise the priority of water and sanitation issues at local, national, and international levels. It will encourage global, regional, and national institutions and organizations to promote best practices and approaches aligned with U.S. interests. It will promote the sharing of best practices and lessons learned and support partner country efforts to peacefully resolve potential conflicts over shared waters.

Strengthening Partnerships, Intergovernmental Organizations, and the International Community: In addition to a focus on private sector engagement, the U.S. government will leverage U.S. investments and promote local leadership through a wide range of partnerships with the international and multilateral community, U.S. and partner country civil society, private sector stakeholders, and other major donors.

Agency-Specific Plans

Many U.S. government agencies and departments engage internationally on water security and sanitation and apply the principles and approaches outlined in this strategy. While some of these agencies and departments have resources appropriated for international work, others advance their missions with domestic resources or are supported by other agencies or third parties to implement activities in support of this strategy.

Agency-specific plans in Annex A describe the activities and approaches identified by each U.S. agency to contribute to the success of this strategy. Likewise, agency-specific plans are a primary means of operationalizing the White House Action Plan on Global Water Security. The Interagency Water Working Group, jointly led by the Department of State and USAID, will provide coordination across the contributing agencies for ongoing planning and implementation of this strategy.

High-Priority Countries

Pursuant to Section 136 of the Foreign Assistance Act of 1961 (FAA), as amended by the Water for the World Act of 2014, the U.S. government will focus its efforts under this strategy on those countries and geographic areas where the needs are greatest and where engagement can best protect U.S. national security interests.

Annex B provides a detailed description of the process to identify High-Priority Countries (HPCs) in accordance with the Act and a list of countries designated for the period October 1, 2022 – September 30, 2023. The U.S. government will implement this strategy in HPCs in accordance with country-specific plans, which will become available as digital appendices to this strategy and on GlobalWaters.org. Federal departments and agencies may determine additional geographic priorities as detailed in Annex A.

In implementing this strategy, the U.S. government recognizes that water insecurity is a challenge that does not recognize political borders. Therefore, work in HPCs will have benefits beyond just those countries. At the same time, efforts under this strategy will not be restricted to HPCs, and regional implications will be considered.

ANNEX A: AGENCY-SPECIFIC PLANS

U.S. Department of State

U.S. Agency for International Development

Centers for Disease Control and Prevention

International Trade Administration

Millennium Challenge Corporation

National Aeronautics and Space Administration

U.S. Army Corps of Engineers

U.S. Department of Agriculture

U.S. Department of Commerce National Oceanic and Atmospheric Administration

U.S. Department of Defense

U.S. Environmental Protection Agency

U.S. Geological Survey

U.S. International Development Finance Corporation

U.S. Trade and Development Agency

U.S. Department of State Plan

Introduction

As the U.S. government's lead foreign policy institution, the Department of State's mission is to shape and sustain a peaceful, prosperous, just, and democratic world, and foster conditions for stability and progress for the benefit of the American people and people everywhere. This includes responsibility for the continuous supervision and general direction of economic and other assistance programs under §622 of the Foreign Assistance Act of 1961 (22 U.S.C. §2382), and coordination of all U.S. foreign assistance, with certain limited exceptions, under 1523 of the Foreign Affairs Reform and Restructuring Act of 1998 (22 USC §6593).

The Department of State views water security as an issue of national security. Its Washington-based officials and those based at embassies and missions worldwide engage with foreign governments and in international fora to promote policies and initiatives to improve global water security. The Department of State also works closely with the National Security Council (NSC) on both the development and implementation of the NSC's 2022 Action Plan for Water Security.

Within the Department of State, the Bureau of Oceans and International Environmental and Scientific Affairs (OES) coordinates the development of U.S. policies and positions through the Department of State Special Advisor for Water Resources and with support from the Interagency Water Working Group (IWWG). OES also works with a host of other bureaus and offices within the Department of State, including regional bureaus, which manage U.S. bilateral and multilateral relationships, and functional bureaus, which manage U.S. foreign policies on specific subject matters. For example, the Bureau of International Organization Affairs (IO) works closely with OES, the Special Presidential Envoy for Climate, the Bureau of Economic and Business Affairs, and the Bureau for Energy and Natural Resources, among others, to advance national interests and priorities, including environment and water, through multilateral diplomacy. IO develops and implements U.S. policy in the United Nations (UN) system and a range of other multilateral organizations through six diplomatic missions in Geneva, Montreal, Nairobi, New York, Rome, and Vienna.

Contribution to the Global Water Strategy

Strategic Objective 1: Strengthen Water and Sanitation Sector Governance, Financing, Institutions, and Markets

The Department of State's diplomatic toolkit is key to ensuring a systematic approach to water and sanitation sector governance. Globally, the Department of State seeks to improve the effectiveness and efficiency of regional and international organizations working on water and issues related to water through direct engagement, dialogue, and partnerships. Alongside interagency partners, the Department of State works to promote sound policies and best practices to foster a supportive enabling environment for water and sanitation investments, including a level playing field for U.S. businesses. In the multilateral context, IO, with its missions in Geneva and Nairobi, works closely with UN-Water and the UN Environmental Program to address water-related issues. U.S. diplomats working in embassies abroad engage with foreign governments on water and sanitation governance issues, provide technical assistance and expertise, and connect U.S. water sector companies to opportunities in-country. In Washington, the IWWG also serves as an engagement opportunity for Commerce, Treasury, and other finance agencies.

Program Examples: The OES-supported U.S. Water Partnership's Water Smart Engagements (WiSE) pairs cities within the Association of Southeast Asian Nations (ASEAN) region with U.S. cities, water districts, and utilities for the purpose of collective capacity building. The overall purpose of the program is three-fold: first, to increase water security in ASEAN partner cities through sustainable water management solutions; second, to establish long-term relationships between ASEAN and U.S. utilities to foster communication and build capacity; and third, to increase the exchange of services, goods, science, and technology.

Strategic Objective 2: Increase Equitable Access to Safe, Sustainable, and Climate-Resilient Drinking Water and Sanitation Services and Adoption of Key Hygiene Behaviors

Ensuring universal access to climate-resilient drinking water and sanitation requires a significant investment of political will at all levels. The Department of State strives to raise the profile of water, sanitation, and hygiene; prioritize water in national development plans; and increase technical capacity in the water sector, especially in regard to water, sanitation, and hygiene objectives. To this end, the Department of State utilizes the full breadth of its global assets and relationships to advocate in diplomatic fora.

Program Examples: The Department of State partners with the U.S. Department of Interior to support the Ambassador's Water Experts Program, which pairs U.S. subject-matter experts with partner countries to improve water, sanitation, and hygiene capacity. From 2022-2027, the Department of State plans to increase water-focused projects as preferred targets for Ambassador discretionary spending projects. The Department of State also will encourage U.S. embassies in priority countries and regions to raise water, sanitation, and hygiene issues in engagement with host countries. The Department of State will move beyond engagement with the ministries of water and raise water, sanitation, and hygiene issues with more influential ministries such as the ministries of interior, finance, and health.

Strategic Objective 3: Improve Climate-Resilient Conservation and Management of Freshwater Resources and Associated Ecosystems

Meeting the challenge of climate-driven water stress will require a substantial change in how the world manages and protects its freshwater resources, already strained from environmental degradation and mismanagement. A resilient approach to sustainable water management can provide a key pathway for meeting economic, environmental, and food security goals while conserving biodiversity and protecting ecosystem function. The Department of State leans heavily on U.S. domestic experience and technical knowledge to support innovative policies, strengthened institutions, and the development of next-generation infrastructure to meet the complex interrelated challenges of the future. Internally, the Department of State works to facilitate the exchange of information among its own officers and diplomats on the role of water and water as a pathway for improving climate security and resilience. Externally, the

Department of State fosters inclusive, data-driven water planning to ensure water systems and water decisions are moving toward climate resilience.

Program Examples: Through the OES-supported HydroSPHERE (Sustainable Planning for Healthy Ecosystems and Resilient Environments) program, the U.S. Army Corps of Engineers (USACE) strengthens water governance by strengthening regulatory mechanisms; increasing the sustainability of infrastructure design, planning, construction, operations, and maintenance; developing public consultation in the national planning processes; and other considerations such as groundwater and green infrastructure. Engagements under this program include bilateral and multilateral capacity building from subject matter expert exchanges, senior leader engagements, technical assistance and consultation, training and workshops, humanitarian assistance construction, and sharing of best practices, tools, strategy, plans, policy development and implementation. A flagship program of the Mekong-U.S. Partnership, the Mekong Water Data Initiative works to strengthen the capacity of Mekong countries to collect, analyze, and manage water and water-related data and information to reduce water-related risks, improve regional responses to environmental emergencies, and promote sustainable economic development across the water, food, energy, and environmental nexus.

Strategic Objective 4: Anticipate and Reduce Conflict and Fragility Related to Water

Availability of water is a key foundation for strong and stable societies. Conflicts over water can exacerbate existing political or societal tensions and undermine national security. The Department of State marshals interagency expertise to encourage partner countries to build the political will to prioritize sound water management and to build their capacity for cooperative management of shared waters. The Department of State works to reduce the role of water as a contributor to conflict through the work of its overseas missions, including under the Global Fragility Act, implemented by the Bureau for Conflict and Stabilization Operations.

Program Examples: The International Union for the Conservation of Nature (IUCN) Shared Waters Cooperation Facility is a global convening platform that brings together diverse partners, inter-governmental organizations, and technical and knowledge partners to coordinate and ensure more effective, consistent, and accelerated support for transboundary water cooperation. The purpose of the Facility is to: empower cooperation on shared waters and support stakeholders to navigate and negotiate agreements for water resource protection and management; provide and source appropriate expertise that can facilitate dialogue and joint action where cooperation has slowed or stopped; and respond to the needs of stakeholders through matching services to problem-solving, connecting solutions, and supporting learning to improve performance.

Approach

The Department of State achieves results through leveraging the breadth of its agency's work in Washington and abroad to facilitate cooperation through diplomatic engagement; financial investment; policy guidance; public diplomacy; and coordinating the U.S. interagency through the management of the IWWG. The IWWG, which meets every month, is the main coordination mechanism for international water efforts for the U.S. interagency. The Department of State's

Special Advisor for Water Resources serves as the lead for these efforts, representing the department in diplomatic engagements and facilitating implementation across the department's bureaus and overseas missions.

Principles

The below guiding principles reflect broad and long-standing U.S. values. They reflect the U.S. domestic approach to water management, including successes and failures. The Department of State has found that meeting the needs of communities in a sustainable, resilient manner requires the full consideration of the political, societal, environmental dynamics that are enumerated in these principles. Likewise, water offers a pathway for the Department of State to engage with international partners on fundamental issues of power, politics, and human rights. Overcoming water challenges therefore can also lead to significant political and economic dividends, including more representative and responsive governments, strengthened agriculture and food systems, stronger regional cooperation, and reduced risks of conflict.

2022 Guiding Principles

- Work through and strengthen global, national, and local systems
- Focus on meeting the needs of marginalized and underserved people, communities, and those in vulnerable situations
- Leverage data, research, learning, and innovation
- Incorporate resilience across all aspects of the strategy

Geographic Focus

The Department of State uses specific criteria to determine high-priority areas for its work. Although water stress impacts all regions, specific basins and watersheds will be considered individually and on a rolling basis, based on the continued availability of resources. Considerations may include, but are not limited to, political opportunity and risk, alignment with parallel U.S. strategies and initiatives, partnership support and growth, and priorities and capacity at U.S. missions abroad.

Resource Implications

The Department of State leverages a wide range of resources, including foreign assistance funds, technical assistance and policy guidance programming, personnel, multilateral negotiations, stakeholder engagement, education, and public diplomacy. Through cooperation with other agencies, the Department of State leverages its resources to catalyze additional investments and build partnerships that achieve meaningful outcomes. Department of State programs and partnerships build political will and strengthen the capacity of regional institutions to manage shared waters cooperatively; to improve global best practices for greater food, energy, and environmental benefits; and to raise the priority of safe drinking water and sanitation among governments. Specific activities, such as targeted capacity building and exchange of best practices, highlight U.S. approaches and technologies while demonstrating the benefits of cooperation.

Assumptions

As climate-related risks grow, the Department of State's ability to implement programs abroad could be stymied by negative climate impacts, such as drought, flood, or other natural disasters. The ongoing COVID-19 pandemic also poses a challenge. In an agency where relationships are key, limited in-person work makes achieving strategic objectives more difficult. In this plan, the Department of State assumes some delays and challenges due to both factors.

U.S. Agency for International Development Plan

USAID Agency Plan Contents

Introduction and Agency Context   **1**

USAID Contributions to the Global Water Strategy and Alignment with other USG Policies and Strategies   **1**

Figure 1: Global Water Strategy Results Framework and Principles with USAID Intermediate Results   4

Figure 2: USAID Targets for the Global Water Strategy period (2022-2027)   5

Strategic Objective 1: Strengthen Water and Sanitation Sector Governance, Financing, Institutions, and Markets

Box 1. Inclusive approaches to advance water security   6

IR 1.1 Develop, strengthen, and implement inclusive laws, policies, and regulations   6

IR 1.2 Effectively mobilize and target public and private financing   7

IR 1.3 Improve the capacity and performance of regional, national, and sub-national institutions   8

IR 1.4 Advance transparency, accountability, equity, and efficiency through participatory, data-driven decision-making   9

Strategic Objective 2: Increase Equitable Access to Safe, Sustainable, and Climate-Resilient Drinking Water and Sanitation Services and the Adoption of Hygiene Practices

IR 2.1 Increase area-wide access to safe, equitable, and affordable sanitation services   10

IR 2.2 Increase access to equitable, safe, reliable, and affordable drinking water services   11

IR 2.3 Improve performance and climate resilience of water and sanitation service providers   12

IR 2.4 Increase adoption of key hygiene practices   13

Strategic Objective 3: Improve Climate-Resilient Conservation and Management of Freshwater Resources and Associated Ecosystems

IR 3.1 Allocate and use water resources more equitably and efficiently   15

IR 3.2 Enhance reliability and quality of water resources through watershed management, including protection, restoration, and nature-based solutions   16

IR 3.3 Improve the climate resilience of water resources management (WRM)   16

Strategic Objective 4: Anticipate and Reduce Conflict and Fragility Related to Water   **17**

IR 4.1 Strengthen capacity to predict, prepare for, and adapt to shocks impacting water and sanitation systems in fragile settings   17

IR 4.2 Address humanitarian water, sanitation, and hygiene needs   18

IR 4.3 Strengthen cooperation and reduce conflict over water   19

IR 4.4 Strengthen coherence across humanitarian, development, and peacebuilding approaches to water and sanitation programming   20

USAID Approaches and Commitments to Mainstreaming Global Water Strategy Operational Principles   **20**

GWS Principle 1: Work through and strengthen global, national, and local systems   20

GWS Principle 2: Focus on meeting the needs of marginalized and underserved people and communities, and those in vulnerable situations   21

GWS Principle 3: Leverage data, research, learning, and innovation   21

GWS Principle 4: Incorporate resilience across all aspects of this strategy   22

Implementing across the USAID Program Cycle   **22**

Designation of High-Priority and Strategic Priority Countries and Regions   22

High-Priority Countries   22

Strategic Priority Countries   22

Strategic Planning   23

Reflecting the Global Water Strategy in Regional and Country Development Cooperation   23

Strategies Developing Individualized Plans for High-Priority Countries   24

Program and Activity Design and Implementation   24

Design Objectives   24

Monitoring, Evaluation, Learning (MEL) and Research   25

Additional Resources and Processes to Support Strategic Programs and Activities   26

Programmatic Budgeting and Resources   27

Roles and Responsibilities   **27**

Global Water Coordinator   27

Water Leadership Council   28

Water and Sanitation Technical Working Group   28

Mission Water and Sanitation Leads   28

U.S. GLOBAL WATER STRATEGY 2022-2027
USAID Plan

Introduction and Agency Context

As the U.S. government's (USG) principal leader, coordinator, and provider of international development assistance, the U.S. Agency for International Development (USAID) advances national security and economic prosperity while demonstrating American values and goodwill abroad. In partnership with the Department of State, USAID co-leads the USG's efforts to implement the Water for the World Act and the U.S. Global Water Strategy (GWS).

USAID's investments in water security under the GWS are critical to advancing health, prosperity, stability, and resilience in the places they are needed most.

During the five-year implementation of the first U.S. GWS (2017-2022), USAID exceeded its targets to provide 15 million people with access to safe drinking water and eight million people with access to sanitation. The Agency recognizes that investments need to stretch further and contribute to transformational and lasting changes. This plan outlines how USAID will scale its impact under the second GWS (2022-2027) by supporting partner countries and communities to:

1. Redouble efforts to close gaps in access to water and sanitation, and hygiene services and products that are sustainable, climate-resilient, and equitable;

2. Build a lasting and strong enabling environment for water security by strengthening institutions and mobilizing financing so that local systems can accelerate progress toward universal access to services;

3. Reduce water stress and build resilience to climate change and other shocks and stressors by expanding the scope and impact of water resources management (WRM) investments and working to conserve and restore watersheds; and

4. Address water-related drivers of and vulnerabilities to conflict and instability by improving linkages and coherence between humanitarian, development, and peacebuilding efforts and improving preparedness for water-related shocks and stressors in fragile contexts.

USAID Contributions to the Global Water Strategy and Alignment with Other U.S. Government Policies and Strategies

This plan details the approach USAID will take to achieve Intermediate Results (IRs) that contribute to each of the four strategic objectives (SOs) of the GWS and the White House Action Plan on Global Water Security (Figure 1). This plan also describes tangible actions the Agency will take to implement and integrate GWS principles across its programming.

To achieve universal access to water, sanitation, and hygiene for all people, healthcare facilities, and schools by 2030, as targeted by Sustainable Development Goal 6 (SDG 6), the pace of progress to extend services globally must at least quadruple, and must accelerate by between four and seven times in the countries where USAID works. Even without the climate crisis, pandemics, and other global shocks, achieving this will require transformational change. It is increasingly clear that the global community needs to focus on strengthening local systems that can provide lasting water security instead of providing quick fixes. Lasting solutions require coordinated, multilateral actions and multi-stakeholder partnerships and strong local leadership and political will.

U.S. GLOBAL WATER STRATEGY 2022-2027
USAID Plan

USAID recognizes that the priorities outlined in the GWS and the White House Action Plan on Global Water Security are more ambitious than any single organization can tackle. As USAID continues to work toward a water-secure world, the Agency's contributions are shifting away from building infrastructure and directly providing services to working with our partners to build and strengthen resilient and equitable systems – including work to strengthen policy and legal frameworks, institutions, and financing systems and markets.

To reflect these shifts, USAID has set ambitious targets that reflect our efforts to directly increase the number of people reached with services, and has also set new types of targets that capture how USAID intends to strengthen essential systems by mobilizing financing and improving institutional governance and performance (Figure 2). Our targets reflect only the impacts of our work that are directly attributable to our programs over the strategy period. Progress against these targets will be monitored primarily through Standard Foreign Assistance Indicators under HL 8.1-8.5. However, the ambition reflected in the targets around financing and governance is that these results contribute to a ripple effect as additional financing, strong institutions, and effective governance continue to accelerate progress toward universal access and a water-secure world.

Thus, during the five years of this strategy (2022-2027), USAID will:

- Work with partner countries to **directly provide 22 million people with sustainable drinking water services** and **22 million people with sustainable sanitation services.**[6] For the first time, the Agency has set equal targets for drinking water and sanitation services to elevate sanitation, which has been historically deprioritized. USAID will pursue these targets across stable and fragile contexts, and at least half of the total number of people reached will be gaining first-time access to basic services.[7]

- Recognizing that one of the biggest barriers to accelerating access to WASH services and overall water security is the financing gap, USAID will directly **mobilize $1 billion dollars of financing for water security, resilient watersheds, sanitation, and hygiene, beyond direct USAID investments.** USAID's ambitious financing target reflects the critical need to more effectively leverage our investments with other sources of funding.[8]

---

[6]This target reflects only the number of people whose increased access to water, sanitation, and hygiene is directly attributable to USG assistance. It does not incorporate increases in access that USAID has contributed to indirectly. While this captures only a portion of USAID's anticipated overall impact, it demonstrates a level of ambition that has grown much faster than the funding levels for water programs. Between the start of the 2017 strategy period and the start of this 2022 strategy period, the Congressional appropriation for water has increased from $400 million to $475 million, or 19 percent. However, USAID's ambition is increasing beyond this. The direct access target for water services has increased by 47 percent (15 million people to 22 million) and the sanitation target has increased by 175 percent (8 million people to 22 million). New targets for finance and institutional strengthening demonstrate USAID's commitment to contributing to larger impacts. The indirect impact on access to water, sanitation, and hygiene services resulting from USAID activities around governance and finance would be reasonably expected to contribute to a much greater number of people gaining access over time.

[7]Increasing support to extending first-time access to basic services aligns with USAID's priority under this plan to focus on ensuring economically or otherwise marginalized people are not left behind. However, extending sustainable basic services to those who have never had it before is not equivalent to comprehensively advancing equity and inclusive approaches in water and sanitation programming. In other words, while the commitment to focus on reaching those who have not had access to basic water and sanitation contributes to USAID's commitment to equity and inclusive development, it does not represent the entirety of our approach (see also Box 1 and our approach to GWS Principle 2).

[8]Note that this financing target is $1 billion dollars to water and sanitation services and water resources management over the five years of this GWS. USAID made a separate commitment under the PREPARE initiative to leverage $1 billion dollars over ten years for climate-resilient water and sanitation services.

- Mobilizing additional financing will only be possible if sector institutions are positioned to attract and invest those funds effectively. USAID will work with partner countries **to measurably improve the performance of more than 1,000 water and sanitation institutions** across at least 30 countries, including all High-Priority Countries (HPCs). Strengthening and improving the performance of institutions, such as local, national, or regional water or sanitation authorities, regulators, service providers, and civil society organizations, is foundational for improving governance, building local capacity, and enabling local systems that lead to lasting change.

In addition to results that can be attributed directly to USG assistance, through our investments in governance, finance, and the broader enabling environment, USAID aims to contribute to the acceleration of progress that is required to reach universal access in the places where we work, estimated to be at least a four-fold increase and significantly more depending on context. USAID will undertake efforts during this five-year period to track and better understand these broader and indirect impacts on the sector in countries where the Agency invests in water security, sanitation, and hygiene. These efforts will be supported by the use of custom indicators and other approaches outlined in USAID's Water and Sanitation Indicator Handbook. The targets and approaches laid out in this plan contribute to and align with government-wide and USAID agency-specific strategies, policies, and mandates, as described in Annex A-1.

Figure 1: Global Water Strategy Results Framework and Principles with USAID Intermediate Results

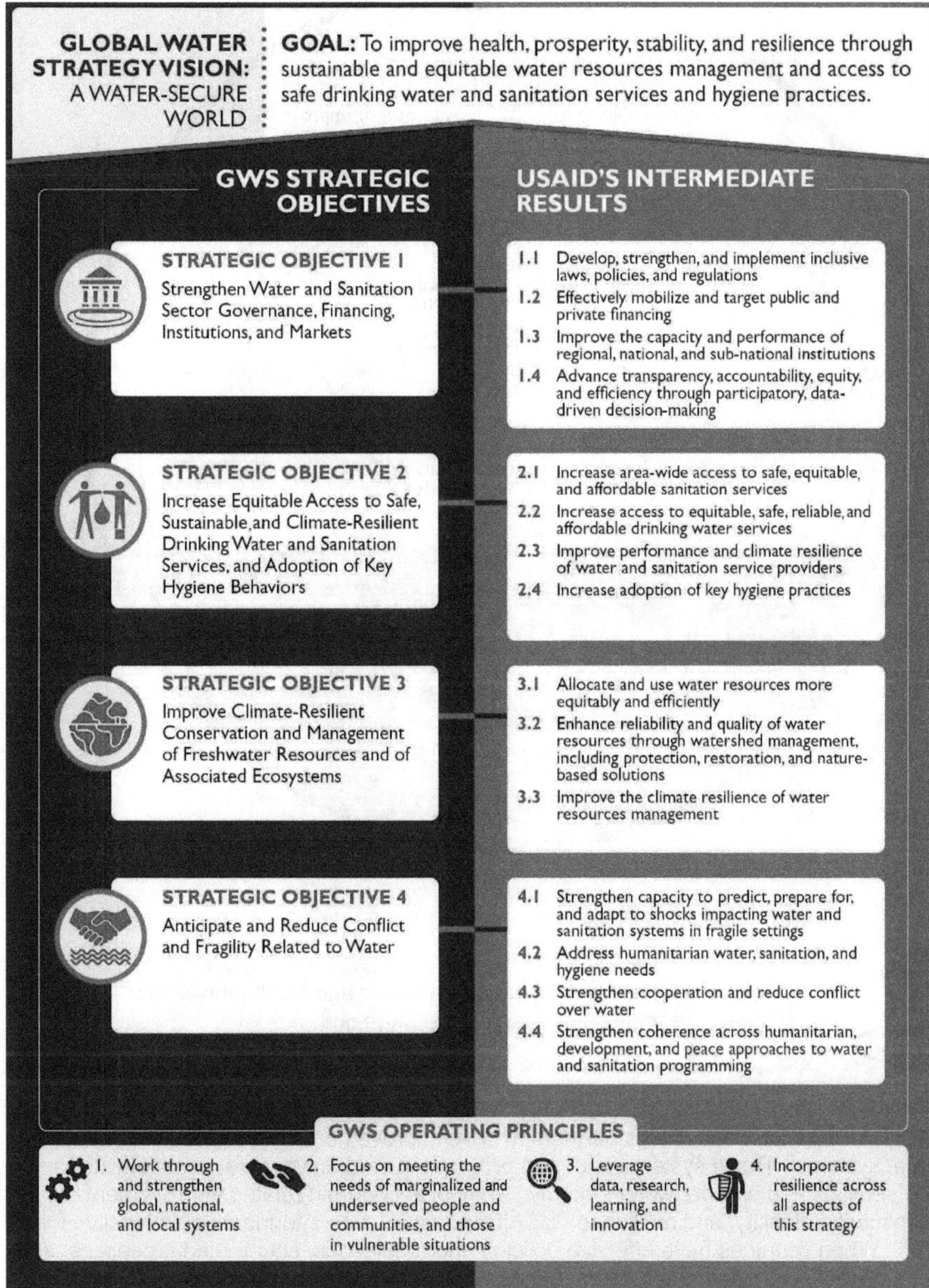

GLOBAL WATER STRATEGY VISION: A WATER-SECURE WORLD

GOAL: To improve health, prosperity, stability, and resilience through sustainable and equitable water resources management and access to safe drinking water and sanitation services and hygiene practices.

GWS STRATEGIC OBJECTIVES

USAID'S INTERMEDIATE RESULTS

STRATEGIC OBJECTIVE 1
Strengthen Water and Sanitation Sector Governance, Financing, Institutions, and Markets

1.1 Develop, strengthen, and implement inclusive laws, policies, and regulations
1.2 Effectively mobilize and target public and private financing
1.3 Improve the capacity and performance of regional, national, and sub-national institutions
1.4 Advance transparency, accountability, equity, and efficiency through participatory, data-driven decision-making

STRATEGIC OBJECTIVE 2
Increase Equitable Access to Safe, Sustainable, and Climate-Resilient Drinking Water and Sanitation Services, and Adoption of Key Hygiene Behaviors

2.1 Increase area-wide access to safe, equitable, and affordable sanitation services
2.2 Increase access to equitable, safe, reliable, and affordable drinking water services
2.3 Improve performance and climate resilience of water and sanitation service providers
2.4 Increase adoption of key hygiene practices

STRATEGIC OBJECTIVE 3
Improve Climate-Resilient Conservation and Management of Freshwater Resources and of Associated Ecosystems

3.1 Allocate and use water resources more equitably and efficiently
3.2 Enhance reliability and quality of water resources through watershed management, including protection, restoration, and nature-based solutions
3.3 Improve the climate resilience of water resources management

STRATEGIC OBJECTIVE 4
Anticipate and Reduce Conflict and Fragility Related to Water

4.1 Strengthen capacity to predict, prepare for, and adapt to shocks impacting water and sanitation systems in fragile settings
4.2 Address humanitarian water, sanitation, and hygiene needs
4.3 Strengthen cooperation and reduce conflict over water
4.4 Strengthen coherence across humanitarian, development, and peace approaches to water and sanitation programming

GWS OPERATING PRINCIPLES

1. Work through and strengthen global, national, and local systems
2. Focus on meeting the needs of marginalized and underserved people and communities, and those in vulnerable situations
3. Leverage data, research, learning, and innovation
4. Incorporate resilience across all aspects of this strategy

Figure 2: USAID Targets for the Global Water Strategy Period (2022-2027)

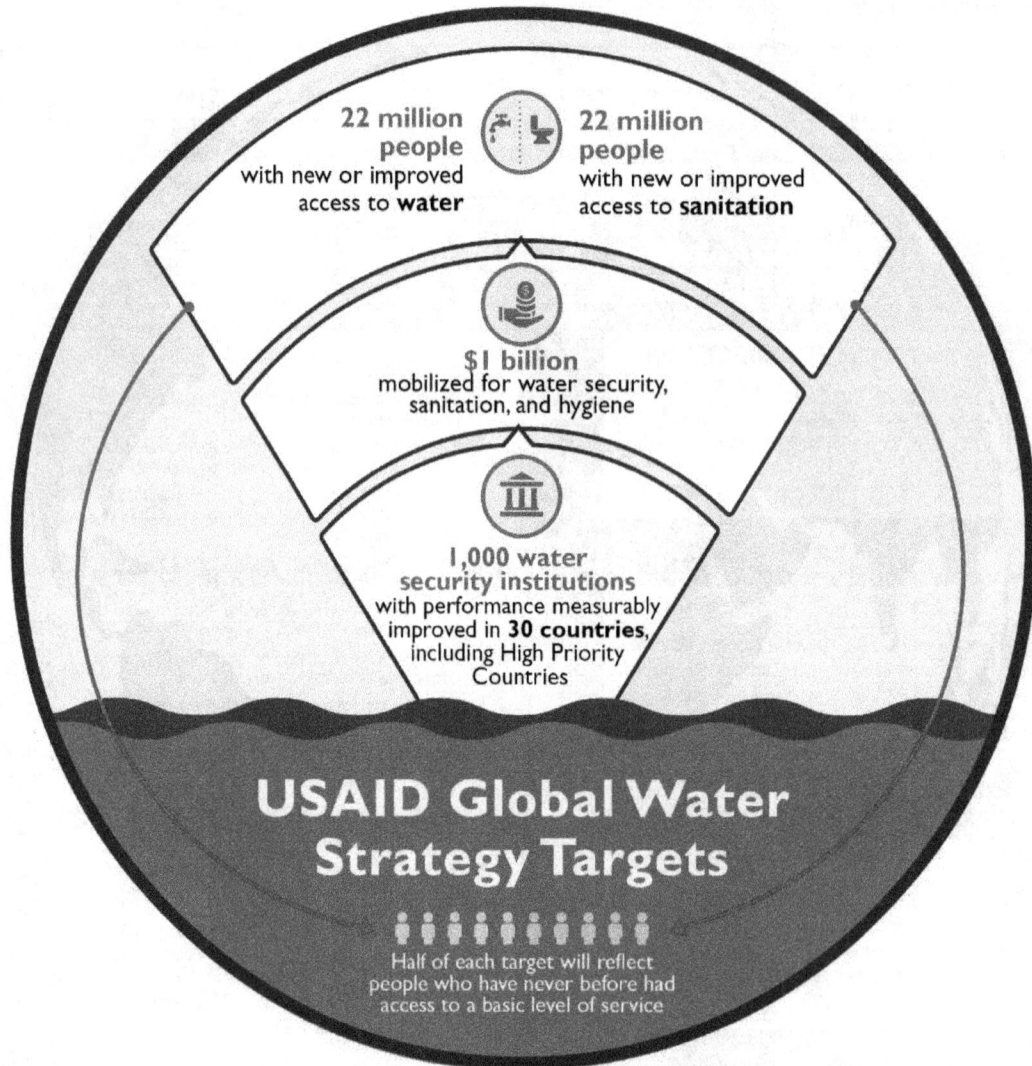

22 million people with new or improved access to **water**

22 million people with new or improved access to **sanitation**

$1 billion mobilized for water security, sanitation, and hygiene

1,000 water security institutions with performance measurably improved in **30 countries**, including High Priority Countries

USAID Global Water Strategy Targets

Half of each target will reflect people who have never before had access to a basic level of service

Strategic Objective 1: Strengthen Water and Sanitation Sector Governance, Financing, Institutions, and Markets

Under SO 1, USAID will work with its partners to drive improvements in governance, finance, and market systems. These systems underpin progress toward universal access to water and sanitation services and broader water security. They also underpin related improvements in health, prosperity, stability, and resilience. USAID views SO 1 as a foundational objective for all other SOs. When countries have effective policies and regulations, country-led processes, and institutions capable of implementing them, they attract investment from both domestic and external sources. A cycle of increased capacity and performance, greater investor confidence, and increased sector finance ultimately accelerates coverage of water and sanitation services.

Through this SO, USAID also seeks to reduce corruption and advance accountability and efficiency. While doing so, USAID will emphasize equitable, inclusive, and sustainable outcomes in alignment with: *GWS Principle 1 - Work through and strengthen global, national, and local systems*; *GWS Principle 2 - Focus on meeting the needs of marginalized and underserved people and communities, and those in vulnerable situations* (see also Box 1); and *GWS Principle 4 – Incorporate Resilience across All Aspects of this Strategy.*

Box 1. Inclusive approaches to advance water security

USAID takes an inclusive development approach in its water security, sanitation, and hygiene investments. This means considering the reasons an individual, household, or community may be unserved or underserved or struggle to access sanitation and water resources. It also recognizes that some people are marginalized by virtue of their membership in specific groups, identities, or for historical or other contextual reasons. These groups include but are not limited to: (1) children in adversity, (2) women and girls, (3) persons with disabilities, (4) lesbian, gay, bisexual, transgender, queer, and intersex (LGBTQI+) people, (5) Indigenous Peoples, (6) youth and older persons, and (7) those experiencing discrimination or marginalization based on their racial, ethnic, or religious identity.[xvi]

USAID also targets approaches to promote and ensure water security for people in vulnerable situations. Vulnerable situations can arise from climate change, state fragility, violence, internal displacement, migration or nomadism, natural disasters, insecure housing, residence in informal settlements, lack of legal identity or tenure, and other challenges that create barriers to water security and sanitation regardless of a person's level or type of marginalization. Each underserved or marginalized person or community, and those in vulnerable situations, may face other intersecting barriers to water security, including but not limited to legal status, poverty, geography, education, social status, profession, and other social, economic, and political factors.

See more about USAID's approaches under Principle 2 and Annex A-1 Policy Linkages.

The success of USAID's efforts to implement SO 1 will be monitored primarily through the following Standard Foreign Assistance Indicators:

- HL.8.3-3: Number of water and sanitation institutions strengthened to manage water resources or improve water supply and sanitation services as a result of USG assistance, disaggregated by "first strengthened this year" and by "Institution Primary Focus (Drinking Water, Sanitation, Water Resources Management)".

- HL.8.4: Value of new funding mobilized to the water and sanitation sectors as a result of USG assistance, disaggregated by factors such as source of funding and funding type.

To achieve SO 1, USAID will implement four interrelated IRs as described below.

IR 1.1 Develop, strengthen, and implement inclusive laws, policies, and regulations

More effective and inclusive laws, policies, and regulations are fundamental to transforming the sector both in terms of sustainably increasing access to water, sanitation, and hygiene services and products for all, and ensuring water resources and watersheds are sustainably managed.[9]

---

[9] For laws, policies, and regulations to be inclusive, they need to reflect the priorities of diverse stakeholders in their language and implementation, and specifically target equitable and inclusive outcomes.

Under this IR, USAID will work with local, national, and regional institutions, including traditional or customary systems, to develop and improve laws, regulations, policies, plans, coordination, and enforcement. The Agency will do this in ways that are inclusive and actionable, enable private sector participation where appropriate, and promote transparency, public participation, and accountability. USAID will seek to enable long-term strategic planning and carefully balance competing objectives when supporting the strengthening of policies and regulations. This includes advising on setting user fees, rate structures, and subsidies so that services are affordable to users while at the same time ensuring long-term financial viability of public or private service providers.

Illustrative activities include:

- Support host governments in the coordinated development, implementation, and enforcement of inclusive, climate-resilient sector laws, regulations, strategies, policies, standards, and institutional and regulatory frameworks to improve water, sanitation, hygiene, and WRM at the international, national, and subnational levels.

- Support processes such as Joint Sector Reviews and other multi-stakeholder platforms or binding institutional frameworks that strengthen coordination and accountability among government and local civil society actors and stakeholders.

- Empower underserved and marginalized communities to inform governance processes.

- Provide technical assistance to help structure and implement economic incentives, such as targeted subsidies to expand services more equitably and rapidly.

- Conduct tailored and on-demand assessments, such as enabling environment assessments, political economy analyses, and market assessments, to better understand power dynamics and perceptions, accountability gaps, conditions, specific barriers to equity, and incentives that shape behaviors and decision-making.

Additional implementation guidance to support this IR can be found in the WASH Governance USAID technical brief and other technical briefs in the USAID Water and Development technical series.

IR 1.2 Effectively mobilize and target public and private financing

Financing needed for delivering sustainable water management and universal access to water, sanitation, and hygiene vastly exceeds current levels of investment.[xvii] Most countries invest less than 0.5 percent of gross domestic product in the sector, well short of the Sanitation and Water for All target of five percent of the national budget.[xviii]

Mobilizing additional domestic resources will require better tracking and monitoring of current public investments and increased advocacy for additional domestic resources such as increased tax revenues and user-fee reforms. Moreover, it is estimated that only 15 percent of water and sanitation utilities in low-income countries (LICs) can cover operations and maintenance costs with user fees, a primary source of revenue.[xix] Service providers become more competitive for additional funding, both private and public, when they first collect existing user fees more efficiently and comprehensively. Finally, expanding access to new and creative sources of funding for the sector, including blended financing mechanisms, revolving funds, repayable commercial finance, and climate funds, is critical to meet growing financial needs of service providers and other sector institutions.

Under this IR, USAID will aim to increase the effectiveness of current funding and mobilize additional public and private funds to expand access to water, sanitation, and hygiene services and products and improve management of water resources.

Illustrative activities include:

- Support the development and implementation of local and national government sector financing plans to maximize and better target existing public funding and mobilize additional funds from domestic public and private resources and user fees.

- Advise water and sanitation service providers on how to strengthen business viability and creditworthiness to unlock access to public and commercial finance including leveraging digital tools and other opportunities to improve efficient collection of user fees.

- Partner closely with multilateral development banks and other development finance institutions (such as the U.S. Development Finance Corporation, national and local banks and investor platforms), to develop, modify, and/or expand financial products and instruments (e.g., using catalytic capital to leverage private finance for blended financing mechanisms, developing payment for environmental services mechanisms, etc.) that help meet demand from service providers and other actors.

- Facilitate increased access to loans, grants, and other financial products for households and water, sanitation, and hygiene businesses and institutions.

- Strengthen public budgeting and financing.

- Provide technical assistance to help structure private sector partnerships to improve management of water resources and reduce collective water-related risks.

Additional implementation guidance to support this IR can be found in the Financing Water and Sanitation Services USAID technical brief and other technical briefs in the USAID Water and Development technical series.

IR 1.3 Improve the capacity and performance of regional, national, and subnational institutions

Fostering a strong and diverse set of public and private institutions is critical to rapidly, equitably, and sustainably extending and improving water, sanitation, and hygiene services and effectively managing water resources across different geographic scales.[10]

Under this IR, USAID will work with local communities and governments to identify needed technical assistance and other investments that foster resilient improvements in the performance of diverse sector institutions and actors. This includes improving the performance of private sector entities, facilitating knowledge and technology transfer, and expanding market-based approaches. Critical elements of improved performance of institutions include advances in diverse human resources and staff capacity, business and operational planning, management systems and technologies, clarification of roles and responsibilities within decision-making

---

[10]Such institutions can include governments and associated regulators, international or regional entities with legal authority, civil society organizations, transboundary or other basin authorities, traditional or customary governance or management entities, and formal or informal public and private water and sanitation service providers from across the water and sanitation value chain (see also SO 2).

processes, occupational health and safety, integration of data and evidence into routine planning, and communication with and responsiveness to customers and other stakeholders.

Illustrative activities include:

- Support human capacity development for professionals in government, the private sector, and civil society through training, mentoring programs, and twinning arrangements that connect institutions to transfer expertise and share best practices.

- Promote human resource systems that recognize, award, and elevate outstanding performers based on clear and transparent metrics.

- Facilitate market entry for private sector partners, particularly local entities, to develop and deploy water, sanitation, and hygiene products and services that contribute to long-term equitable expansion of services, and support small and medium water and sanitation enterprises to succeed.

- Provide technical assistance and capacity-building support to national and subnational government departments or agencies for improved annual budget process, contracting, and supervision.

- Strengthen institutional capacities to integrate adaptive management principles, strategies, and technologies into planning frameworks to ensure effective governance in the face of uncertainty, including from climate change.

- Support diversification of the sector's workforce by providing formal training, professionalization, certification, and mentoring to youth, underemployed groups, and civil society organizations.

Additional implementation guidance to support this IR can be found in the WASH Governance USAID technical brief, as well as other technical briefs in the USAID Water and Development technical series.

IR 1.4 Advance transparency, accountability, equity, and efficiency through participatory, data-driven decision-making

Expanding access to water, sanitation, and hygiene services equitably and sustainably requires participation beyond formal ministries and service providers to also include individuals, communities, and advocacy organizations, especially those led by and for underserved or marginalized groups and communities.

To advance IR 1.4, USAID will support transparent and accountable decision-making by working to ensure that a diversity of constituents, especially marginalized and underrepresented people, can safely and meaningfully participate in the sector, including in sector research, data collection, and decision-making. USAID will support civil society organizations in holding governments accountable, including in locations where partnership with governments is limited due to corruption and government-sanctioned conflict. To ensure that efforts to increase equity are evidence-based, USAID will invest in collection, analysis, sharing, and increased use of evidence and data to improve decision-making, transparency, and adaptive management.

Illustrative activities include:

- Support national statistical agencies and multi-stakeholder learning platforms in the collection of, standardization of, and public access to, high-quality data about services and water uses and resources, such as water quality information, fecal waste treatment performance, key performance indicators for service providers, service coverage, and budgets.

- Facilitate, recognize, support, and incorporate community data collection and mapping as a basis of participatory planning, budgeting, and co-design or development of basic services at the community and city levels.

- Promote and work to institutionalize stakeholders' participation, capacity, and leadership, especially women, underrepresented groups, and civil society focused on advocacy and accountability.

- Support institutions, processes, and tools for government and service providers to hear directly from and be held accountable by users and other stakeholders, such as by establishing within laws and regulations transparent and accessible complaint and appeals processes and forums.

Strategic Objective 2: Increase Equitable Access to Safe, Sustainable, and Climate-Resilient Drinking Water and Sanitation Services and the Adoption of Hygiene Practices

Under SO 2, USAID seeks to partner with local government and public and private sector service providers to expand access to safe, affordable, reliable, and climate-resilient water, sanitation, and hygiene services and products across entire cities, districts, or counties, including in institutional settings like schools and healthcare facilities. USAID will emphasize working through local systems (see GWS Principle 1) and align with work under SO 1 to improve the performance of service providers and market actors along water, sanitation, and hygiene value chains. USAID will also support state-of-the-art social and behavior change approaches that concurrently address individual, structural, and social factors to increase the adoption and sustained practice of key hygiene behaviors, including for menstrual health. This SO is critical to advancing health, dignity, and educational and economic opportunity, especially for women and children.

SO 2 will be monitored primarily through Standard Foreign Assistance Drinking Water Indicators (under HL.8.1) and Sanitation and Hygiene Indicators (under HL.8.2). These indicators monitor access to both basic and safely managed services and number of people whose service quality improved as a result of USG assistance. They also monitor the number of schools and health facilities gaining access to basic water or sanitation services as a result of USG assistance. HL 8.2-5 monitors households with soap and water at a handwashing station on premise and is critical to tracking the success of IR 2.4. All of the individual-level indicators must be disaggregated by sex (male, female) and, as outlined in Principle 2, should be increasingly disaggregated by marginalized group over the lifetime of this strategy.

IR 2.1 Increase area-wide[11] access to safe, equitable, and affordable sanitation services

New evidence indicates that health, nutrition, early childhood development, and other positive downstream outcomes of better sanitation are only achieved when improvements occur across

---

[11]Area-wide refers to the population within an entire geographical area, typically aligned with governmental administrative boundaries, such as a district, province, or city.

entire geographies—that is, when most households across an entire community or neighborhood or larger area have access to improved sanitation.[xx] To maximize these benefits, USAID will focus its work at area-wide scales, aligning with local administrative units, and market systems. USAID will also seek to support outcomes across the entire sanitation service chain and shift its programming toward more contextual, layered, and sequenced approaches that are necessary to ensure coverage across the diversity of geographic, market, and population demographics that exist at these scales. The drive toward increasing the number of people with access needs to be paired with a focus on reducing persistent inequities in access.[xxi] Overall progress has sometimes been made at the expense of increasing inequality, which in turn is strongly linked to and exacerbated by compounding vulnerabilities such as poverty, lack of land or housing tenure, disability, gender, and climate change (see also Box 1).

Under IR 2.1, USAID will simultaneously seek to strengthen local governance, financing, markets, and service delivery systems (see also SO 1) while supporting targeted efforts to progressively reach the most underserved and marginalized people and those in vulnerable situations.

Illustrative activities include:

- Support programs that seek to promote increased adoption of sanitation and hygiene products, services, and behaviors.

- Promote achievement of safely managed sanitation that considers climate risks and includes infrastructure and services along the full sanitation service chain, including fecal sludge management.

- Conduct sanitation market assessments to understand and strengthen markets and improve the viability of sanitation businesses.

- Support capacity building and workforce development to improve sanitation leadership, planning, design, construction, monitoring, and operations and maintenance.

- Support efforts to specifically reach poor and marginalized people and those in vulnerable situations through targeted programs and subsidies, and through engagement of non-traditional and underrepresented stakeholders in decision-making.

- Strengthen access to sanitation in institutional settings, including healthcare facilities and schools, by working with government partners to integrate sanitation and hygiene into overall planning and monitoring processes, particularly as part of education, quality of care, and infection prevention and control efforts.

Additional implementation guidance to support this IR can be found in the Urban and Rural Sanitation Services and WASH in HCF for Quality Health Systems USAID technical briefs, as well as other technical briefs in the USAID Water and Development technical series.

IR 2.2 Increase access to equitable, safe, reliable, and affordable drinking water services

Increasing access to safe, reliable drinking water services brings proven health and economic benefits to households, communities, and nations.[xxii] Reducing the distance and eliminating physical barriers between home and point of access to water has also been linked to increased use and safety of water, improved nutritional status, decreased child mortality, improved mental health, and reduced bodily injury from water fetching.[xxiii] Provision of an improved drinking water service on premise with high water quality has been shown to reduce diarrhea risk by up to 52 percent.[xxiv] More proximate access to drinking water, including on premise, is particularly important for gender equity in low-income households, where women and children tend to bear disproportionate responsibility for water collection, contributing to significant time poverty.[xxv] Effective, safe, and independent access for women, girls, and persons with disabilities is critical to minimizing risks of gender-based violence, including exploitation for water or sanitation and hygiene supplies, and deteriorating health or hygiene.

Under IR 2.2, USAID will work with partners to increase access to both basic and safely managed water services, decreasing the distance to point of access, improving water quality and service reliability at the point of use, and increasing affordability and access to water on premise. To reach more people with progressively higher quality and more affordable services, USAID will promote area-wide approaches by working with public and private sector partners to strengthen and/or develop service delivery models, noting that approaches may differ between urban and rural areas. While USAID will support construction of new infrastructure, it will focus heavily on improved service provider operations and maintenance and asset management practices to ensure the sustainability of infrastructure investments and services.

Illustrative activities include:

- Construct or rehabilitate water infrastructure and provide technical assistance in the planning, financing, and implementation of infrastructure and related services, with emphasis on reliable operations and maintenance arrangements, and equity and accessibility of the resulting services.

- Support the development, innovation, and scaling of viable drinking water service models, including those that focus on or ensure at least basic services for low-income and marginalized populations.

- Strengthen water safety, quality, and quantity monitoring systems and increase the capacity of service providers to undertake routine water quality testing.

- Strengthen access to water in institutional settings, including healthcare facilities and schools, by working with government partners to integrate water, sanitation, and hygiene into overall planning and monitoring processes, particularly as part of education, quality of care, and infection prevention and control efforts.

- Conduct analyses of the impacts of conflict on access to water and sanitation services for marginalized groups and support use of the findings in water service planning, infrastructure development, and scale-up efforts.

Additional implementation guidance to support this IR can be found in the Urban and Rural Water Services and WASH in HCF for Quality Health Systems USAID technical briefs, as well as other technical briefs in the USAID Water and Development technical series.

IR 2.3 Improve performance and climate resilience of water and sanitation service providers

Building on SO 1, under this IR USAID will support improving the performance of all types of service providers with a focus on promoting principles of commercial operations while recognizing that many service providers will continue to need public finance to expand and enhance services equitably. To address the growing climate risks to safe and sustainable sanitation and drinking water services, USAID will also strengthen service providers' ability to understand and develop the adaptive capacities necessary to respond to complexity and uncertainty, such as from rapid urbanization, climate change, and other threats. Working toward low-emissions, resilient water and sanitation services is a high priority for USAID, fundamental to protecting development gains and national and economic security, and to supporting climate adaptation amongst those who are most vulnerable to climate impacts.

Illustrative activities include:

- Help service providers to improve their business, financing, and investment planning to drive improvements in revenue collection, creditworthiness, and other key performance indicators and benchmarks, leveraging digital tools where appropriate.

- Strengthen the operational and technical capacity of service providers including encouraging the diversification of the workforce and promotion of women and other marginalized populations in leadership roles.

- Support service provider networking, trade organizations, and learning mechanisms such as twinning, to improve performance and accountability, including for small and informal providers.

- Support utilities to reduce non-revenue water, implement demand-management systems, improve energy efficiency, and utilize renewable forms of electricity where appropriate.

- Provide technical assistance to service providers and their regulators to conduct vulnerability assessments and climate risk analyses, to use the assessments to develop standards for improving the resilience of new and existing household and municipal infrastructure, and to plan for business continuity and emergency response.

- Support service providers to engage more proactively in groundwater and watershed management to improve water quality and availability at the point of supply.

Additional implementation guidance to support this IR can be found in the USAID Water Security, Sanitation, and Climate Change technical brief and other technical briefs in the USAID Water and Development technical series.

IR 2.4 Increase adoption of key hygiene practices

Hygiene behaviors and products are essential to health and nutrition and ensure that drinking water and sanitation services and products are used consistently. Under IR 2.4, USAID will work with individuals, communities, and relevant institutions -such as health, water, and education systems - in a locally led, context-driven manner to support the adoption and consistent use of key hygiene practices. Key hygiene practices include, but are not limited to, handwashing with soap, safe household drinking water management, safe food hygiene, safe handling of animal feces, and menstrual health and hygiene. USAID will also work with local partners to support efforts to shift social and gender norms linked to water, sanitation, and hygiene, adjust incentive structures, and work toward improving policies that foster a stronger enabling environment. Finally, USAID will support market-driven approaches to ensure that critical hygiene supplies are manufactured and available locally at affordable prices.

Illustrative activities include:

- Research consumer and household preferences, needs, and barriers to behavior change to strengthen social and behavioral change (SBC) investments and drive market-driven solutions to ensure supplies are affordable and available to all market segments.

- Support institutions to strengthen policy, regulations, and laws that enable market availability and accessibility of multiple product choices to support hygiene and sanitation behaviors.

- Design and implement evidence-based SBC interventions that address key structural, social, and individual behavioral determinants and promote positive social and gender norms.

- Build capacity and knowledge of community health workers, educators, and members of civil society to influence individual and communal behaviors and to endorse norm-shifting interventions.

Additional implementation guidance to support this IR can be found in the USAID Menstrual Health and Hygiene and Social and Behavior Change for Water Security, Sanitation, and Hygiene technical briefs and technical briefs in the USAID Water and Development technical series.

Strategic Objective 3: Improve Climate-Resilient Conservation and Management of Freshwater Resources and Associated Ecosystems

Under SO 3, USAID will invest in watershed conservation and WRM to support partner countries in their efforts to plan for rising water stress so that water is available to support vibrant and healthy communities and cities, sustainable food and energy systems, and healthy ecosystems. Building on investments under SO 1, USAID will focus its implementation of SO 3 on supporting WRM institutions to engrain planning, decision-making, and local actions that are informed by data, that are more inclusive, and that enhance resilience to climate change.

WRM is a multisectoral challenge that involves bringing together and aligning the many different sectors that rely on water and finding sustainable solutions that balance the needs of communities and cities, food and energy systems, and ecosystems. Connecting USAID's work under this SO to other USAID initiatives that advance food security, biodiversity conservation, and climate change adaptation and mitigation efforts will be critical to help ensure the reliability of water for agricultural and food security, economic productivity, ecosystem health, and to meet basic human needs[12] (see Annex A-1 Policy Linkages).

Additional implementation guidance to support all of the IRs under SO 3 can be found in the USAID Water Security, Sanitation, and Climate Change and WRM technical briefs and other technical briefs in the USAID Water and Development technical series.

Investments under SO 3 will be monitored using Standard Indicators under HL. 8.5.

---

[12]USAID's work under this SO aligns with and directly reinforces GFSS IR 6, "Improved Water Resources Management." USAID views WRM to be a joint responsibility between Water for the World Act and GFSS initiatives and funding streams.

IR 3.1 Allocate and use water resources more equitably and efficiently

Effective WRM that supports broad-based economic growth, improves equity, and enhances climate resilience involves understanding available water resources and developing systems or plans to govern how to allocate those resources across users within and between basins. Many critical basins lack comprehensive or efficient water management and allocation plans. Where such plans have been developed, current water supply and quality are rarely sufficient to sustainably meet the competing needs of all users, future changes in demand and climate are not taken into account, or laws and related plans are not being implemented or enforced.

Under IR 3.1, USAID will work to advance more sustainable and holistic use of water resources and more equitable, robust water resource planning and allocation of water across users and ecosystems within or between river basins (including transboundary basins). In doing so, USAID will apply two critical elements of improved, robust, and equitable water resources planning, allocation, and use (see also SO 1). First, USAID will increase engagement of a broad range of stakeholders to identify water-related risks and trade-offs, develop solutions to improve the quantity and quality of available resources (see also Box 1 and IR 3.2), and advance more equitable and efficient allocation. Second, USAID will support improved collection and routine use of hydro-meteorological data, including of surface and groundwater, as the basis for understanding water availability and use at multiple spatial and temporal scales.

Illustrative activities include:

- Support the adoption of policies to enable more equitable, robust, and flexible allocation of water resources, prioritizing water for ecosystems and domestic needs.

- Support the development of policies to protect water quality from threats such as agricultural runoff, untreated wastewater, and industrial and commercial discharges.

- Facilitate the development and implementation of stakeholder-driven water allocation and integrated WRM plans in subnational and transboundary basins, focusing on groups that have faced barriers to participating in these processes.

- Convene and support empowerment of water user groups in water resource planning, with a specific focus on including women, Indigenous Peoples, persons with disabilities, and other underserved and marginalized groups and people in vulnerable situations.

- Support participatory analysis of existing land and water tenure, including customary and traditional rights, practices and systems of Indigenous Peoples and other marginalized groups.

- Strengthen capacities of water user associations, regulatory agencies, and laboratories to routinely collect and utilize data and information on water use, hydrometeorological conditions, and water quality, including through tools and services that leverage remote sensing and hydrological and climate monitoring and modeling.

- Improve the collection of water abstraction and user fees.

- Promote practices and innovations that improve water use efficiency, conservation, and water reuse.

IR 3.2 Enhance reliability and quality of water resources through watershed management, including protection, restoration, and nature-based solutions

Activities that improve water storage and moderate water flow or availability over time greatly enhance the success and resilience of water management strategies. These benefits can also extend to other development objectives, such as biodiversity conservation, improved public health, and greenhouse gas mitigation.

Under IR 3.2, USAID will work within watersheds to promote the design and implementation of actions that enhance water storage and groundwater recharge, improve water quality and reduce the cost of water treatment, restore and maintain river flow and enhance resilience to floods and droughts, and increase water-related ecosystem services. Interventions will be chosen as part of watershed planning processes where there is collaboration across sectors and stakeholders to consider multiple potential goals and the costs-benefits of different types of interventions. Traditional knowledge and Indigenous Peoples' practices are resources for innovative and nature-based solutions. USAID will seek to build partnerships with Indigenous Peoples and to support their leadership in planning and implementing such solutions, including by improving linkages between traditional governance entities and national governments as needed.

Illustrative activities include:

- Support holistic analysis and stakeholder engagement necessary to identify priority watersheds and actions for investment.

- Strengthen natural systems and promote nature-based solutions to maintain ecosystem goods and services, such as soil conservation, upstream reforestation, and wetland restoration and conservation.

- Promote other green infrastructure such as the construction of infiltration ponds, sand dams, and vegetative buffer strips.

- Establish water funds and markets that link downstream water users with upstream landowners to help pay for planning and conservation and restoration efforts.

IR 3.3 Improve the climate resilience of water resources management (WRM)

Under this IR and in conjunction with work under SO 1 and SO 4, USAID will work to increase preparedness and reduce vulnerability to flooding, drought, and other water-related events such as hurricanes and extreme monsoons.

Illustrative activities include:

- Provide technical assistance to national, regional, and local governments to incorporate climate risks into water security policies and plans, including through flood and drought risk assessments and maps.

- Work with communities, local leaders, and institutions to enhance capacity for analyzing and using climate and weather data in decision-making and to manage climate-related risks and uncertainties regarding quality and availability of water resources.

- Pursue local solutions and embrace a participatory approach to identify water security hazards and solutions to improve future water resource availability.

- Support the capacity development of local leaders in effective communications and engagement to empower their communities to participate in policy reform to support climate- and water-related risk management.

Strategic Objective 4: Anticipate and Reduce Conflict and Fragility Related to Water

As the world faces more frequent and intense conflicts, extreme weather events, and climate-related migration, USAID will work in fragile contexts to reduce water-related drivers of and vulnerabilities due to conflict and fragility through a multifaceted approach that focuses on systems preparedness, emergency response when needed, conflict mitigation, and coherence across approaches. Given that conflict and disasters often reveal and reinforce systemic inequalities,[xxvi] USAID will maintain a "do no harm" posture and will focus on those marginalized populations that are at greatest risk. This includes people who experience disproportionate rates of gender-based violence in conflict and disaster settings,[xxvii] displaced populations who face compounding water-related vulnerabilities,[xxviii] and other groups who face discrimination due to underlying power structures that perpetuate the cycle of water-related fragility, conflict, and vulnerability.[xxix]

Additional guidance to support implementation of the IRs under SO 4 can be found in USAID's Humanitarian-Development Coherence in WASH or WRM Programs and Menstrual Health Hygiene technical briefs, as well as other technical briefs in the USAID Water and Development technical series.

Progress under SO 4 should be monitored using a combination of HL.8 and other Standard Indicators and custom indicators to ensure that the cross-cutting nature of this SO is captured. Standard Indicators include "Number of individuals directly utilizing improved water services provided with [Bureau of Humanitarian Affairs] BHA funding" (W29) and "Number of individuals directly utilizing improved sanitation services provided with BHA funding" (W13).

IR 4.1 Strengthen capacity to predict, prepare for, and adapt to shocks impacting water and sanitation systems in fragile settings

In the past 20 years, over half of all natural disasters were floods (44 percent), droughts (6 percent), and other water-related events. Flooding can also accelerate the spread of waterborne diseases, such as cholera, which spreads primarily in areas with inadequate water, sanitation, and hygiene access.[xxx] These disasters led to 1.6 million human deaths, the vast majority (82 percent) of which were in LICs and lower-middle-income countries (LMICs). Water and conflict have a reciprocal relationship; drought can be a predictor of sociopolitical unrest, while conflict can lead to damage to water infrastructure and reductions in access and quality.[xxxi]

And while disasters can and do happen everywhere, focusing on reducing disaster risk in places that face recurrent shocks makes economic sense. Estimates suggest that every dollar spent on social safety-net or resilience-building in areas of recurrent drought will result in three dollars of benefit in terms of avoided asset losses and reduced humanitarian spending.[xxxii]

Under IR 4.1, USAID will address chronic water vulnerabilities and reduce the overall impact of recurrent water shocks and stresses in fragile contexts through a focus on enhancing systems that reduce disaster risk and build local, national, and regional response capacities to confront disasters.

Illustrative activities include:

- Provide technical assistance to national and regional governments and traditional and customary governance institutions to establish and codify emergency response strategies appropriate for disasters, water-related disease outbreaks, and armed conflict.

- Work to improve national and regional meteorological services for seasonal climate forecasting as part of improving preparedness, especially for drought.

- Empower community members and civil society, including organizations led by and for underserved and marginalized groups, to participate in disaster preparedness planning and to assess the capacity of hygiene practices and water and sanitation infrastructure to mitigate known hazards.

- Invest in systems and planning to support the distribution and storage of commodities procurement and stockpiling for disaster preparedness.

- Work with other donors to proactively mobilize funding to ensure water and sanitation service delivery to displaced populations.

- Engage national governments, financial institutions and insurance providers, and micro-entrepreneurs to develop weather-indexed insurance in areas prone to droughts and floods.

IR 4.2 Address humanitarian water, sanitation, and hygiene needs

Emergency water, sanitation, and hygiene needs are already acute. One hundred-twenty million people were in need of WASH-related humanitarian assistance in 2021, while a 75 percent funding gap[13] existed and around 40 percent of those humanitarian WASH needs went unmet.[xxxiii] In 2022, basic emergency water and sanitation needs are projected to increase in 56 of 58 countries identified as "fragile contexts"[xxxiv] while rising conflict is set to exacerbate the scope of those emergency needs globally.[xxxv] Safe drinking water, sanitation, and hygiene interventions in particular are essential for curbing the impacts of infectious disease outbreaks and mitigating disproportionate impacts of crises, such as gender-based violence, on marginalized groups. Meanwhile, WASH remains one of the most chronically underfunded sectors within humanitarian assistance.[xxxvi]

Under this IR, USAID will address global humanitarian water and sanitation needs through water supply, sanitation, hygiene promotion, environmental health, menstrual health and hygiene, and WASH non-food items interventions to reduce morbidity and mortality resulting from shocks or displacement and safeguard dignity and personal security.[xxxvii] These interventions focus on meeting the immediate needs of populations, particularly in consultation with vulnerable and displaced groups.

Illustrative activities include:

- Repair critical water and sanitation systems that have been damaged by disaster or conflict, considering groundwater resource sustainability and multiple water uses (including livestock and agriculture).

---

[13]Humanitarian WRM funding is not tracked globally so the funding gap is likely large but unknown.

- Distribute WASH non-food items (e.g., water treatment, menstrual hygiene supplies, hygiene kits), and train people on their effective use and key hygiene practices.

- Support staffing for critical humanitarian WASH coordination roles.

- Incorporate protection and gender-based violence prevention and safeguarding principles into humanitarian water and sanitation programs.

IR 4.3 Strengthen cooperation and reduce conflict over water

Conflict and cooperation over water co-exist and are interrelated; it is rare that a given interaction over water can be categorized purely as a "conflict" or as "cooperation."[xxxviii] Even when it is not the direct cause of conflict, water resource disputes can exacerbate existing fragility and disputes between social groups, regions, or nation-states.[xxxix] Meanwhile, at local levels, there have been many cases where cooperation over water has been maintained despite periodic, minor conflicts, such as between herders and farmers,[xl] and cooperation has the potential to create better outcomes through water activities that use systems approaches.[xli]

Under IR 4.3, USAID will seek to promote and apply systematic tools to reduce potential water-related conflicts and/or to minimize how water programming could inadvertently trigger conflict.

Illustrative activities include:

- Leverage tools, such as the Conflict Assessment Framework, the Water and Conflict Toolkit, and the Land and Conflict Toolkit, to integrate conflict analysis and do no harm principles in program design and to identify linkages between water, conflict, and proposed USAID activities.[14]

- Support the collection and use of shared data for decision-making across countries and stakeholder groups.

- Bring together and build the capacity of and trust between stakeholders, including relevant regional organizations, to establish or improve governance processes in order to prevent the escalation of a water conflict.

- Promote cross-learning processes for transboundary water cooperation among different stakeholders.

- Work with key local stakeholders in flashpoint areas to prevent water-based conflicts from escalating and spreading, with a focus on supporting conflict-sensitive approaches to water management through training and enhanced technical expertise.

IR 4.4 Strengthen coherence across humanitarian, development, and peacebuilding approaches to water and sanitation programming

Fragile contexts demand a nuanced approach to sector programming, requiring simultaneous, coordinated efforts across humanitarian, development, and peacebuilding assistance.[xlii] This coordination can take the form of complementary, yet separate programs in the same geographic area, or fully integrated programs that blend different humanitarian, development, and peacebuilding approaches within a single activity. Integrated approaches require more conscious program development efforts and engagement with host country governments, including through joint analysis and design. At their core, both approaches require that development actors focus on strengthening systems at all levels to facilitate emergency

---

[14]New versions of these toolkits are forthcoming.

response when needed, and that humanitarian actors create an enabling environment for long-term development when addressing shocks.[xliii]

Under IR 4.4, USAID seeks to align water and sanitation technical approaches across humanitarian, development, and peacebuilding programming in fragile contexts through coordinated planning, analysis, and measurement.

Illustrative activities include:

- Joint planning and analysis across relevant stakeholders to identify entry points and inform initial designs, annual work planning, and collaborating, learning, and adapting (CLA) activities, including through stakeholder mapping, stakeholder analysis, scenario planning, gender and protection analyses, and conflict analysis.

- Developing locally owned, shared metrics for success in watershed, water, sanitation, and hygiene programming across humanitarian, development, and peace actors operating within overlapping geographic zones.

- Monitoring support at the onset and throughout the duration of shocks, including the implementation of baseline and endline surveys to measure outcomes.

- Support complementary programming, such as engaging the private sector or using market-based approaches, while ensuring the basic needs of people in vulnerable situations are met.

USAID Approaches and Commitments to Mainstreaming Global Water Strategy Operational Principles

The GWS operational principles are core values that will guide USAID water and sanitation investments across all SOs and the USAID Program Cycle. Under this plan, USAID will also employ specific means to track alignment with and progress against each principle, as described below and in the Program Cycle section.[15] Recommended standard and custom indicators for each principle below are included in USAID's Water and Sanitation Indicator Handbook.

GWS Principle 1: Work through and strengthen global, national, and local systems

Achieving a water-secure world requires working at multiple levels, engaging and strengthening global, national, and community systems. Adopting a systems approach requires intentional efforts to understand layered contexts and the stakeholders that impact water and sanitation outcomes in specific settings, including civil society, the private sector, and traditional or customary governance structures. To ensure uptake of Principle 1, USAID water, sanitation, hygiene, and WRM programs are all expected to contribute to SO 1. The programs are strongly encouraged to employ systems analysis tools, such as those described in the Agency's WASH Governance technical brief, to identify specific problems and ways to address them across the Program Cycle. Programs should invest in capacity development of key stakeholders at the local and national levels to support local actors to design and lead sustainable solutions and to

---

[15]Where needed, USAID will update internal results reporting tools such as key issue requirements and indicator definitions, and/or provide additional technical guidance to support the broad and meaningful uptake of these principles. These are referenced in the Program Cycle section of this Plan.

drive increased sector financing. This includes engaging civil society in USAID activity design, implementation, monitoring, and evaluation, and alignment with the New Partnerships Initiative.

GWS Principle 2: Focus on meeting the needs of marginalized and underserved people and communities, and those in vulnerable situations

Achieving universal access to water, sanitation, and hygiene means leaving no one behind, yet many people in both stable and fragile contexts face barriers to accessing services and may lack agency in WRM based on factors including, but not limited to, sexual orientation, gender identity, gender expression, sex characteristics, race, ethnicity, religion, class, disability, legal status, geography, and/or age (see also Box 1). Principle 2 is embedded in approaches, priorities, and illustrative examples across all IRs in this plan. USAID will ensure application of Principle 2 across its Program Cycle by ensuring that program and activity design reflects and responds to specific contexts, experiences, traditional knowledge, and barriers to equity and inclusive outcomes. Social and inclusive development analytical tools, including mandatory gender analyses, will be used and findings incorporated across activity design, implementation, and monitoring.

To improve the inclusivity and equity of water, sanitation, and hygiene outcomes, USAID will partner with underserved and marginalized people and communities and with civil society organizations led by and for them and people in vulnerable situations. This includes, for example, seeking free, prior, and informed consent; supporting capacity development to accomplish locally defined goals; empowerment for meaningful participation in activity and policy design, implementation, monitoring, and research; honoring local and traditional knowledge and systems; and using inclusive and accessible communications tools and social and behavior change approaches.

USAID will disaggregate program data, to the extent feasible, to improve understanding of progress, opportunities, and lessons learned in reaching those who are hardest to reach (see Implementing across the USAID Program Cycle). Increased USAID reporting of indicator disaggregates, including disaggregation by sex and, where possible, gender, and by other marginalized demographic factors (e.g., youth, persons with disabilities, or Indigenous Peoples) will better guide the Agency's impact. USAID will also seek to improve USAID activity reporting on Performance Progress Reporting Key Issues such as Menstrual Health and Hygiene and Gender to incentivize and scale up the scope and impact of cross-cutting inclusive development activities within water, sanitation, hygiene, and WRM programs. These efforts are key to the Agency realizing its 2022-2027 target of ensuring half of all people directly reached with water and sanitation services are receiving first-time access.

GWS Principle 3: Leverage data, research, learning, and innovation

Data, research, learning, and innovation are core to USAID's ability to ensure Agency water and sanitation investments are impactful, climate-resilient, equitable, and sustainable. Principle 3 hinges on (1) maximizing CLA practices in water and sanitation programs; (2) rigorously evaluating Agency water, sanitation, hygiene, and WRM programs to draw lessons for future implementation, including through the lens of inclusive development; (3) supporting data generation, use, and sharing by partner countries and local actors; and (4) investing in research, including existing research mechanisms across the Agency, that answers key questions about implementation and global trends for decision makers, as described in the Agency's Water for the World Research Agenda.

GWS Principle 4: Incorporate resilience across all aspects of this strategy

Shocks and stressors – ranging from floods, droughts, pandemics, and conflict to job loss or catastrophic family illness – have strong links to water, sanitation, and hygiene, as reflected across USAID's IRs. To support household and community resilience, USAID will incorporate adaptive approaches for responding to shocks and stressors into water and sanitation activity design, implementation, monitoring, and evaluation as appropriate. Increasing ambition to analyze and incorporate programming design elements to respond to risks identified through Climate Risk Management screening will also be important to realize Principle 4 across the USAID water and sanitation portfolio, as will capturing results from such activities using standard climate change adaptation indicators, as appropriate.

Implementing Across the USAID Program Cycle

USAID strives to ensure that water, sanitation, hygiene, and WRM programming is strategically focused for maximum impact and sustainability. The sections below provide high-level guidance and guidelines for developing water and sanitation programming through the USAID Program Cycle, including strategic planning, project and activity design and implementation, developing activity budgets and resourcing, and monitoring, evaluation, and learning (MEL). While the GWS guides investment across development and emergency programming, there can be exceptions to the guidance in this section when programming outside the development assistance framework.

Designation of High-Priority and Strategic Priority Countries and Regions

High-Priority Countries

Selection of HPCs is described in Annex B of the overall GWS. Officially, HPC designations are made annually; however, USAID intends to retain a consistent set of HPCs for the five-year life of this strategy. A current list of countries designated as high priority under the Water for the World Act can be found on GlobalWaters.org.

Per the Water for the World Act, HPCs will be the primary recipients of USG official water and sanitation development assistance. Under this plan, HPCs are also required to meet requirements for strategic planning, staffing, programming, and monitoring, as described in Box 2.

Strategic Priority Countries

Strategic Priority Countries (SPCs) are places where USAID anticipates substantial and long-term investment in water and sanitation due to a combination of strategic considerations and development needs. SPCs are not eligible for designation as HPCs because they do not have high or medium-high needs as defined by the Needs Index, but they are critical countries for USAID engagement on water and sanitation for reasons including national security and other geopolitical considerations, and water scarcity and stress. SPCs are also designated annually, given high levels of sustained investment, and expected to deliver impactful programming aligned with this strategy and plan. Best practices and specific requirements for SPCs around strategic planning, programming, and monitoring, are outlined in the sections below. A current list of countries designated as SPCs can also be found on GlobalWaters.org.

Box 2. What does it mean to be a Water for the World High-Priority Country Mission?

Given the significant investment and focus on water, sanitation, hygiene, and WRM needs and opportunities in HPCs, the designation comes with an expectation of greater reach and impact of HPC programming. To support this ambition, USAID/Washington will prioritize HPC Missions for technical support, capacity development, and access to centrally funded field support activities. To ensure rapid and effective implementation of this strategy and the Water for the World Act, HPCs are also subject to several expectations and requirements:

Field staffing: The Water for the World Act requires HPC Missions to identify a lead subject matter expert who can help deliver impactful programming.

Country Plans and Strategic Planning: The Water for the World Act requires HPCs to develop plans that include budgets and are evidence-based and results-oriented to deliver on Agency water security objectives. HPC Missions are strongly encouraged to link this plan with other high-level planning processes and documents, such as Regional and Country Development Cooperation Strategies (R/CDCSs).

MEL and Research: To demonstrate the impact in HPCs, HPC Missions must report annually on standard water and sanitation indicators, including at least one standard indicator that captures results under SOs 1 and 2, as well as disaggregates noted below. HPCs are encouraged to contribute to sector learning as relevant locally via investment in closing knowledge gaps as laid out in the Water for the World Act Research Agenda. With the support of USAID/Washington, HPC Missions will be required to conduct monitoring and assessments to enable USAID to understand how they are delivering on the approaches in the GWS, including technical pivots, operational principles, and contributions to accelerating the pace of progress needed to meet SDG 6.

Details and additional recommended implementation practices for HPCs and other Operating Units (OUs) are included below throughout the Program Cycle and Roles and Responsibilities sections of this plan.

Strategic Planning

Reflecting the Global Water Strategy in Regional and Country Development Cooperation Strategies

Strategic planning for most bilateral and regional USAID missions is reflected in Country Development Cooperation Strategies (CDCSs) and R/CDCSs. The CDCS provides a guide for the subsequent design of projects and/or activities to operationalize specific results. Thus, careful consideration of water and sanitation challenges and objectives is important when developing R/CDCSs. Per Automated Directives System (ADS) 201.3.2.6, missions must align their R/CDCSs to Agency and Interagency strategies/policies. HPCs and SPCs that develop new or make significant revisions to existing R/CDCSs during this five-year plan period are strongly encouraged to reflect water, sanitation, and hygiene priorities in their R/CDCSs, such as through establishing an IR or sub-IR linking directly to the GWS and USAID Agency Plan Results Framework. USAID/Washington will engage HPC and SPC Missions during phase one of the R/CDCS development process and work collaboratively to align R/CDCSs and strategy priorities and objectives. Other missions programming Water for the World Act-authorized development assistance are also encouraged to consider this approach.

Developing Individualized Plans for High-Priority Countries

The Water for the World Act specifies that USAID develop individualized plans for designated HPCs as part of an appendix to the Agency's strategy requirements. The aim of HPC plans is to ensure that the country-level strategies for implementation advance and align with the revised strategy, agency-specific plans, and the CDCS. HPC plans are to be "costed, evidence-based, and results-oriented." USAID will work to update and develop new HPC plans on a rolling basis and post plans to GlobalWaters.org.

Program and Activity Design and Implementation

Design Objectives

USAID projects and activities funded under the Water for the World Act must be designed so that they contribute to one or more of the SOs of the GWS and the associated IRs of this plan, while meaningfully reflecting GWS principles. The design process is a unique opportunity to understand local contexts and systems and leverage the best available data and evidence to:

Maximize Impact: Effective design ensures that the greatest number of targeted people and systems receive the greatest possible benefit, delivering the most transformative impact possible using the available resources.

Improve Equity: Marginalized populations are often the hardest to reach, but investing the resources needed to do so is a core principle of the GWS and the Water for the World Act (Principle 2). As such, all water, sanitation, hygiene, and WRM program and activity designs should include analysis to better understand and explicitly reduce inequity in water and sanitation access and benefits. This includes inequities between wealthier and poorer populations that typically have access to lower quality and more expensive services and inequities due to sexual orientation, gender identity, gender expression, sex characteristics, level of physical and mental ability, Indigenous status, age, religion, ethnicity, location, and other factors (see Box 1). Water, sanitation, hygiene, and WRM project and activity designs should also ensure that such marginalized populations have input into the design process and participate meaningfully and safely in activity implementation and monitoring and research. In addition to the mandatory Gender Analysis, USAID recommends the use of Inclusive Development Analysis and Social Impact Assessment tools.

Increase the Likelihood of Durable Results and Localization: Ensuring that the impact achieved can be sustained and expanded in the future requires taking a systems approach that, among other things, includes increasing local ownership and investing in local capacities. To ensure alignment with Principle 1, water, sanitation, hygiene, and WRM programs should all at a minimum contribute to SO 1 and work to strengthen governance, financing, institutions, and markets and the relationships between them. Local knowledge and expertise should be emphasized where possible to better tailor programming to specific social, cultural, political, economic, and environmental contexts and support empowerment and sustainability.

Meeting these design elements can be facilitated by analysis to identify the best geographic and/or thematic fit for programming and the approaches that will best suit the target context.

Monitoring, Evaluation, Learning, and Research

Continuous MEL and research help USAID tell our story and provide opportunities for improved CLA. Five key components capture this plan's approach to MEL and research.

Standardized Reporting: Reporting on high-level indicators and their disaggregates is essential for capturing progress toward the goal and SOs of this strategy and form a core part of the water and sanitation MEL framework. Standard indicators are measures that USAID and the Department of State use to collect performance data that can be aggregated globally to help justify requests for funding, understand operational challenges, assess progress, and support external reporting. Standard indicators related to water, sanitation, hygiene, and WRM are found under HL.8 in the Department of State Office of Foreign Assistance's Standardized Program Structure and Definitions. USAID has updated the suite of standard indicators to capture results under each SO, and USAID's Water and Development Indicator Handbook describes how to apply and measure against water, sanitation, hygiene, and WRM standard and custom indicators. To ensure better reporting on related climate standard indicators in support of the USAID Climate Strategy, please refer to the reporting section of the USAID Water Security, Sanitation, and Climate Change Technical Brief.

Every USAID activity utilizing water directive funding must report on at least one standard water and sanitation indicator. To advance the governance and finance priorities laid out in this plan in addition to Principle 1, High-Priority and Strategic Priority OUs must, at a minimum, report on standard water and sanitation indicators that reflect the plan's targets. That means reporting on HL.8.3-3 or HL.8.4-1, in addition to an HL.8.1 and HL.8.2 indicator each year, in addition to other standard indicators as relevant. Other OUs are encouraged to do the same, as appropriate.

Where appropriate, activities should report on standard Global Climate Change indicators to demonstrate progress in transforming the emissions profile and climate vulnerability of water and sanitation systems in alignment with the USAID Climate Strategy and PREPARE Initiative.

To better align with strategy operating principles on equity and resilience and to ensure progress toward meeting the targets in this plan, HPC and SPC OUs must also report on existing key standard indicator disaggregates, including, as appropriate:

- Sex and, where available and appropriate, gender
- Marginalized groups (e.g., as defined in Box 1 and by ADS 201)
- Number of institutions strengthened for the first time
- Funds mobilized for climate-resilient water and sanitation services

Assessment and Monitoring of Programming Pivots, Principles, and Outcomes: USAID will monitor and assess the Agency's progress toward programming and results that reflect the priorities and associated pivots in the GWS and this plan, including progress toward integrating principles and contributing to accelerating access at the rate of change needed to achieve universal access to water and sanitation. HPCs must contribute to these assessments through supplementary monitoring of programming.

Evaluation: Evaluations of certain USAID water, sanitation, hygiene, and WRM programs and activities may meet the evaluation requirements under ADS 201.3.6.5. When OUs undertake evaluations of water and sanitation programs or activities, they are strongly encouraged to incorporate evaluation of the programming priorities, principles, and desired outcomes laid out in the plan, including the efficacy of governance and finance, inclusive development, climate-resilient, and locally led development approaches as they relate to water and sanitation.

Coordinated Research and Learning: To elevate and coordinate research across Water for the World Act activities and programs, USAID has developed a Water for the World Act Research Agenda. Through existing and future research activities, USAID/Washington is committed to investing significant resources over the GWS period to advance knowledge against the gaps identified therein and to maximize the impact of the Agency's investments through the improved evidence base for programming. In alignment with the GWS principle on research and innovation (Principle 3), OUs are encouraged in their activity and programming learning agendas to contribute to at least one of the research gaps identified in the Research Agenda. HPCs and SPCs in particular should work to make meaningful contributions to these broader research objectives.

Data and Information Sharing: Data and information produced by each activity is an essential resource for future learning and programming. As per ADS 579, data should be shared through the Data Development Library and other appropriate government agencies and global data efforts (as appropriate). As per USAID Acquisition Regulation (AIDAR) 752.7005, all program learning documents are also required to be posted to the Development Experience Clearinghouse.

Additional Resources and Processes to Support Strategic Programs and Activities

Technical Guidance: A synthesis of recent global evidence and recommended programming approaches associated with SOs, IRs, and principles and different programming contexts can be found in USAID's Water and Sanitation Technical Brief Series. USAID/Washington will periodically add to and update this technical series with new topic areas and additional evidence or other guidance, as appropriate.

Capacity Building and Training: USAID/Washington will implement a capacity-building strategy for USAID field staff. The Agency will focus on professionalizing the sector backstop within the Agency and offering opportunities for professional development to maximize the Agency's internal capacity to achieve the goals and objectives of the GWS and this plan. Through four interrelated work streams, USAID's capacity-building efforts will seek to promote, improve, and create opportunities for water and sanitation mission staff to apply best practices, systems, processes, and tools to design, manage, learn from, and adapt water and sanitation programs to increase the effectiveness of programming globally. Offerings include a global biannual workshop to facilitate learning and evidence exchange, a focus on enhancements to internal systems and processes (including regularizing Foreign Service National fellowship opportunities, creating career ladders, and pursuing a pilot foreign service backstop), improving internal knowledge management tools, and providing networking support. The Continuous Learning Series provides key knowledge, skills, and abilities required to be an effective leader in water security, sanitation, and hygiene.

Water, Sanitation, Hygiene, and WRM Portfolio Reviews: HPC and SPC Missions are strongly encouraged to undertake a portfolio review annually, either by incorporating it into existing processes such as the mission's preset periodic overall Portfolio Review or as a dedicated exercise. These reviews should be conducted in participation with the Mission

Director, water and sanitation technical staff and other relevant offices, and the Global Water Coordinator or their designee and should be led by mission water and sanitation leads (see Roles and Responsibilities). USAID/Washington will support mission staff who choose to conduct reviews to discuss current and planned investments, project and activity evaluations, mission water directive pipeline, results, challenges, pivots, and tradeoffs that elevate water and sanitation priorities and achieve strategy SOs.

Programmatic Budgeting and Resources

The Water for the World Act authorizes USAID to engage in drinking water and sanitation service provision, hygiene, and WRM in developing countries. Since 2008, annual appropriations acts have provided legal authority for USAID to spend funds on water and sanitation service provision and hygiene promotion. Recent annual appropriations also include subdirectives; since FY 2008, USAID has received a subdirective on water, sanitation, and hygiene in Sub-Saharan Africa; beginning in FY 2015, USAID has received a subdirective on safe latrines. Sector spending must conform to annual appropriations language and Agency-specific policy, including as laid out in this strategy and plan. Technical recommendations on allocation levels and subsequent use of approved levels of water and sanitation directive (e.g., earmark) by the USAID Office of Budget and Resource Management and the Department of State/Office of Foreign Assistance will serve as a guide and input to the formal budget process.

Where outcomes benefit myriad development sectors (e.g., reducing student absenteeism or enhancing government effectiveness) or where gains across other sectors benefit water and sanitation (e.g., enhancing public financial management or agricultural water use efficiency), blended or integrated funding approaches should be considered. In particular, achieving SO 3 and SO 4 is a shared responsibility across numerous appropriations, initiatives, and earmarks. For instance, where watershed conservation or other WRM benefits are not principally linked to the provision of drinking water, USAID will use agriculture, biodiversity, or other funds as appropriate. Similarly, where water investments mitigate escalating conflict between communities, USAID will blend democracy or other funds, as appropriate.

Roles and Responsibilities

Clear organizational roles and responsibilities are necessary to ensure effective implementation of the Agency plan's SOs and that they are in line with its principles. The roughly 80 USAID personnel stationed at headquarters and across the world, and entities listed below, have specific roles related to water, sanitation, hygiene, and WRM programming. All entities below also play a role in coordinating with other sector stakeholders to reduce the risk of duplication and leverage partner priorities.

Global Water Coordinator

The Water for the World Act statutorily requires that the USAID Administrator serve concurrently as, or appoint at the Deputy Assistant Administrator level or higher, a Global Water Coordinator (GWC) to oversee USAID water, sanitation, and hygiene programs, co-lead the implementation and revision of USAID's portion of the GWS with the Department of State, and expand USAID's program capacity in HPCs. In addition, the GWC represents the Agency on issues related to water security, sanitation, and hygiene to Congress and the National Security Council and at external conferences and other events (Public Law 113-289).

Water Leadership Council

The USAID Water Leadership Council (WLC) is an intra-agency coordination platform with representation from all pillar and regional Bureaus at the Deputy Assistant Administrator level. Fundamentally, the WLC elevates the visibility and importance of water and sanitation programming in the Agency, particularly among leadership within USAID's missions. The WLC provides overall leadership and oversight to respond to agency institutional and administrative challenges and opportunities affecting programming. The WLC is chaired by the GWC—positioned within the Bureau of Resilience and Food Security—and deputy chaired by a representative from the Bureau of Global Health. The WLC is composed of representation across both pillar and regional Bureaus engaged in providing resources and technical oversight of water security activities. The key functions of the WLC are to support and coordinate (1) recommendations on budget allocations to the Office of Budget and Resource Management, (2) technical leadership, (3) technical policy guidance, and (4) programmatic oversight of water and sanitation activities and investments within the Agency. These specific tasks stem from ADS Guidance on Leadership Councils.

Water and Sanitation Technical Working Group

The USAID Water and Sanitation Technical Working Group (WS-TWG) is a Washington-based water and sanitation coordination and leadership platform composed of technical staff representing pillar and regional Bureaus and missions with water and sanitation programming. This platform provides a multi-sector structure for collaboration across the extended Washington-based water team in areas such as technical and implementation coordination, strategy development, balancing of equities, and policy recommendations. Ad-hoc sub-working groups may be formed to address specific initiatives, discuss topical challenges, or advise OUs per the request of mission water and sanitation leads and USAID/Washington water and sanitation mission support points of contact (PoCs). The WS-TWG also provides technical guidance and recommendations to the WLC. The Director of the USAID/Washington Center for Water Security, Sanitation, and Hygiene serves as the chair of this group and the liaison between the WS-TWG and WLC.

Mission Water and Sanitation Leads

Mission water and sanitation leads are mission-based subject matter experts who coordinate and lead the mission water and sanitation portfolio and coordinate with USAID/Washington water and sanitation mission support PoCs, U.S. embassy team in the host country, host country officials, and other donors active in advancing water security in the host country. HPC Missions are required by the Water for the World Act to designate water and sanitation leads, but all missions with significant water, sanitation, hygiene, and WRM portfolios are strongly encouraged to establish this role. Especially at HPC and SPC Missions, water and sanitation leads should be full-time staff with strong water, sanitation, hygiene, or WRM sector technical backgrounds. Leads should be at either Office Director or Project Management Specialist level serving within U.S. Direct Hire, U.S. Personal Service Contractor, Third Country National, or Cooperating Country National capacity, and should be responsible for managing the water security portfolio at their missions, including guiding the design, implementation, monitoring and evaluation of water, sanitation, hygiene, and WRM activities, and ensuring alignment with the GWS and this plan. USAID/Washington will support the professional development of water and sanitation leads and other technical staff through its capacity-building strategy (see above).

ANNEX A-1: Policy Linkages

The approaches and objectives in the USAID Plan will help achieve broader development outcomes: health, prosperity, stability, and resilience. In addition to contributing to all three pillars of the whole-of-government White House Water Security Action Plan, this plan has a bidirectional relationship with a number of other government and agency-specific strategies, policies, and mandates, both contributing to their achievement of results and outcomes, while also benefiting from their strategy and programming.

Democracy, Human Rights, Governance, and Stabilization
Efforts to improve water and sanitation sector governance under SO 1 will seek to reduce corruption and advance accountability, contributing to the U.S. Strategy on Countering Corruption and the U.S. Strategy on Democracy, Human Rights, and Governance. Improved sector data and participatory, data-driven, service model innovations, and transparent decision-making will depend upon use of new digital tools to increase equity and efficiency in services, advancing the USAID Digital Strategy. Under both SO 3 and SO 4, the plan advances the U.S. Strategy to Prevent Conflict and Promote Stability as it seeks to anticipate and reduce conflict and fragility related to water, including the promotion of more equitable water allocation.

Environment and Natural Resources Management
Water resources and watershed conservation and restoration investments under SO 3, especially green infrastructure and nature-based solutions, enhance biodiversity conservation, contributing to biodiversity and development outcomes in the USAID Biodiversity Policy. This work also aligns with the cross-sectoral approach of the Agency's Environment and Natural Resources Management Framework, and advances the targets and principles of the USAID Climate Strategy, especially on natural and managed ecosystems, adaptation, and nature-based solutions. The Agency's activities under SO 3 also complement the WRM IRs of the U.S. Global Food Security Strategy.

Gender Equality and Empowerment
The plan's focus on reaching women and girls with access to water, sanitation, hygiene, and menstrual health and hygiene information, services and supplies, and on promoting their participation in water-related decision-making and leadership contributes to the National Strategy on Gender Equity and Equality and the USAID Gender Equality and Women's Empowerment Policy as well as to USAID's Economic Growth Policy by lifting barriers to women's participation in the economy.

Health and Nutrition
Efforts under SO 2 to increase access to safe water and sanitation and the adoption of key hygiene behaviors, coupled with institution strengthening work done under SO 1, align with USAID's Acting on the Call: Preventing Child and Maternal Deaths agenda and advances USAID's Vision for Health Systems Strengthening. Improved management of freshwater resources and associated ecosystems that can ensure the availability of water for agricultural and food security uses, contributes to the goals of the U.S. Global Food Security Strategy and USAID Multi-Sectoral Nutrition Policy.

Inclusive Development
Under SO 2, efforts to create effective, independent, and safe access to water and sanitation services for persons with disabilities; lesbian, gay, bisexual, transgender, queer, and intersex people; and youth will advance the USAID Disability Policy, USAID Policy on Standards for

Accessibility for the Disabled in USAID-Financed Construction, the USAID LGBT Vision for Action, and the USAID Youth in Development Policy. The plan's objective to reach those left behind, particularly under SO 2, will help build strong beginnings for children, contributing to the U.S. government Advancing Protection and Care for Children in Adversity Strategy by improving children's health, nutrition, and ability to access educational opportunities.

Under SO 3, the plan directs USAID programming to strengthen engagement with Indigenous Peoples in water allocation and related planning decisions to safeguard against harm and support their development priorities aligned with the USAID Policy on Promoting the Rights of Indigenous Peoples.

Investments across this plan respond to the Executive Order on Advancing Racial Equity and Support for Underserved Communities through the Federal Government and to the Presidential Memorandum on Advancing the Human Rights of LGBTQI+ Persons Around the World.

Localization
Institution strengthening under SOs 1 and 2 advances the USAID Local Capacity Development Policy, the USAID Private Sector Engagement Policy, and USAID Sustainable Service Delivery in an Increasingly Urbanized World Policy. Diversity, equity, and inclusion underpin a strong and capable water and sanitation workforce and enable USAID to better reach those being left behind, in alignment with the White House Executive Order on Diversity, Equity, Inclusion, and Accessibility in the Federal Workforce, the USAID Diversity, Equity, Inclusion, and Accessibility Strategy, and the USAID Equity Action Plan.

Resilience
Reliable and affordable access to water, sanitation, and hygiene services and products can protect households, communities, and economies facing unpredictable climate, conflict, and health-related shocks and stressors. The approaches laid out in this plan contribute to the USAID Building Resilience to Recurrent Crisis Policy, USAID Climate Strategy, and the USAID Vision for Health System Strengthening.

ANNEX A-2: Glossary

Affordability: Affordability is measured by looking at bill burden: water and wastewater bills should be no more than 4 percent of a household's annual income.

Area-wide: Area-wide refers to the population within an entire geographical area, typically aligned with governmental administrative boundaries, such as a district, province, or city. *Source: Adapted from WaterAid*

Basic drinking water: Drinking water from an improved source, provided collection time is not more than 30 minutes for a round-trip, including queuing time. Improved sources include those that have the potential to deliver safe water by nature of their design and construction, and include: piped water, boreholes or tubewells, protected dug wells, protected springs, rainwater, and packaged or delivered water. Note that basic drinking water for healthcare facilities is defined as water from an improved source that is available on-premises. *Source: WHO/UNICEF Joint Monitoring Programme (JMP): Drinking Water*

Basic sanitation: Use of improved facilities that are not shared with other households, but where excreta is not safely managed. *Source: WHO/UNICEF JMP: Sanitation*

Blended finance: The strategic mobilization of additional capital to match public funds to increase sustainable development in developing countries. *Source: USAID Private Sector Engagement Policy*

Climate resilience: Climate resilience can be generally defined as the capacity of a system to maintain function in the face of stresses imposed by climate change and to adapt the system to be better prepared for future climate impacts. *Source: USAID Climate Strategy 2022-2030*

Ecosystem-based adaptation: A nature-based method for climate change adaptation that can reduce the vulnerability of societies and economies to climate stressors. This includes using nature-based methods to address aspects of water insecurity through strengthening natural systems to maintain the goods and services that ecosystems provide for human development. *Source: Global Eco-based Adaptation Fund*

Ecosystem services: The short- and long-term benefits people obtain from ecosystems. They include: 1) provisioning goods and services, or the production of basic goods such as food, water, fish, fuels, timber, and fiber; 2) regulating services, such as flood protection, purification of air and water, waste absorption, disease control, and weather impact related regulation; 3) cultural services that provide spiritual, aesthetic, and recreational benefits; and 4) supporting services necessary for the production of all other ecosystem services, such as soil formation, production of oxygen, crop pollination, carbon sequestration, photosynthesis, and nutrient cycling. *Source: USAID Biodiversity Policy*

Equity: The consistent and systematic, fair, and just treatment of all individuals, including individuals who belong to marginalized and underrepresented groups that have been denied such treatment. Equity addresses the specific and proportionate needs of certain persons or groups to attain fair and just treatment and outcomes, as opposed to equality, which when used to describe a process, emphasizes the same or equal treatment for all persons or groups regardless of specific circumstances or needs. *Source: USAID Climate Strategy 2022-2030*

31

Enabling environment: The set of interrelated conditions such as legal, governance and monitoring frameworks, political financing and human capital that are able to promote the delivery of WASH services. *Source: UN Water Global Analysis and Assessment of Sanitation and Drinking Water (GLAAS) 2017*

Fecal sludge management: The system for collecting, transporting, and treating fecal sludge from onsite sanitation such as pit latrines and septic tanks. Fecal sludge is made up of human excreta, water, and solid waste that is disposed of in onsite toilets and sanitation systems. Fecal sludge management is required for safely managed sanitation service where a centralized wastewater transport and treatment system is lacking. *Source: Environment and Public Health Organization: Fecal Sludge Management*

Governance: Governance is a concept comprising a set of actors, systems, and processes – including but not limited to government – that determine who gets what type or quality of water, sanitation, or hygiene service, when they get it, how it is paid for, how fast the service is restored if there is an interruption, and who can access water resources. *Source: USAID Water and Development Technical Series: Water Security, Sanitation, and Hygiene Governance*

Gray infrastructure: Engineered structures built with conventional methods, such as conventional steel and concrete drainage and water treatment systems (i.e., pipes, pumps, ditches, storm drains, dams, and detention ponds engineered by people to manage stormwater and drinking water). Conventional treatment systems include energy-intensive water treatment systems and processes such as membranes and reverse osmosis. *Source: Duke Nicholas Institute for Environmental Policy Solutions: Stormwater Management-Gray Infrastructure*

Green infrastructure: Any engineered structure that uses vegetation, soils, and natural processes to manage water and create healthier built environments for people and the natural resources that sustain them. Green infrastructure can range in scale from small-scale technologies such as rain gardens and green roofs to regional planning strategies targeting conservation or restoration of natural landscapes and watersheds. Green infrastructure may be interconnected with existing and planned gray infrastructure to create sustainable infrastructure that can enhance community resilience to disasters and climate change. *Source: U.S. Environmental Protection Agency: What is Green Infrastructure?*

Groundwater: Water found underground in the cracks and spaces in soil, sand, and rock. It is stored in and moves slowly through geologic formations of soil, sand, and rocks called aquifers. It is a preferred source of drinking water as it is often isolated from sources of contamination at the surface. *Source: U.S. Geological Survey: What is Groundwater?*

Groundwater abstraction: The process of taking water from a ground source, either temporarily or permanently. *Source: European Environment Agency: Groundwater Abstraction*

Humanitarian-development-peace coherence: Humanitarian-development-peace coherence aims to promote complementary collaboration across humanitarian, development, and peace actors in pursuit of a common agenda. Its goal is to maximize impact and sustainability of programs across different kinds of assistance and to reduce the need for humanitarian assistance over time. *Source: USAID Programming Considerations for Humanitarian-Development-Peace Coherence*

Improved drinking water: Improved drinking water sources are those that have the potential to deliver safe water by nature of their design and construction, and include piped water, boreholes or tubewells, protected dug wells, protected springs, rainwater, and packaged or delivered water. *Source: WHO/UNICEF JMP: Drinking Water*

Improved sanitation facilities: Improved sanitation facilities hygienically separate excreta from human contact and include: flush/pour flush to piped sewer system, septic tanks, or pit latrines; ventilated improved pit latrines, composting toilets, or pit latrines with slabs. *Source: WHO: Improved Sanitation and Drinking Water-Sources*

Institution: A government, non-government, or parastatal organization with equities or responsibilities in the water and/or sanitation sectors. These institutions may be formal, informal, or customary and include government, civil society, the private sector, and service providers. *Source: UN Environment Program: Institution*

Integrated Water Resources Management: A process that promotes the coordinated development and management of water, land, and related resources. *Source: USAID Water and Development Technical Series: Water Resources Management Technical Brief*

Joint sector review: A process, usually led by the government, which brings together multiple types of stakeholders in regular dialogue to review policy implementation, budget execution, and performance. *Source: USAID Water and Development Technical Series: Water Security, Sanitation, and Hygiene Governance*

Limited sanitation: Households have access to a facility that is considered improved, but that is shared with other households. *Source: WHO/UNICEF JMP: Sanitation*

Limited water service: Households have access to a basic water source, but with collection time greater than 30 minutes round-trip. *Source: WHO/UNICEF JMP: Drinking Water*

Locally led development: The process in which local actors – encompassing individuals, communities, networks, organizations, private entities, and governments – set their own agendas, develop solutions, and bring the capacity, leadership, and resources to make these solutions a reality. *Source: USAID What is Locally-Led Development Fact Sheet*

Local systems: The interconnected sets of actors—governments, civil society, the private sector, universities, individual citizens, and others—that jointly produce a particular development outcome. The "local" in a local system refers to actors in a partner country. As these actors jointly produce an outcome, they are "local" to it. As outcomes may occur at many levels, local systems can be national, provincial, or community-wide in scope. *Source: USAID Local Systems: A Framework for Supporting Sustained Development Report*

Market-based approaches: Market-based approaches use business models and catalyze markets to solve development and humanitarian challenges more sustainably and at scale. A market-based approach can engage low-income people as customers and supply them with products and services they can afford; or, as business associates (suppliers, agents, or distributors), to provide them with improved incomes. When a market-based solution becomes commercially viable, the private sector has a financial incentive to continue and operate it at scale, which increases the sustainability of the intervention, and decreases the need for donor support over time. *Source: USAID Private-Sector Engagement Policy*

Menstrual Health and Hygiene (MHH): Menstrual Health and Hygiene is the ability of women, girls, and transgender and gender non-binary individuals who menstruate ("menstruators" or "individuals who menstruate") to manage their menstrual cycles in a safe, dignified, healthy, and supported manner throughout their lives. *Source: USAID Menstrual Health and Hygiene Technical Brief*

Nature-based solutions: Actions to protect, manage, and restore ecosystems that address societal challenges effectively and adaptively are called nature-based solutions when broadly referring to goals like climate adaptation and mitigation or water and food security. *Source: UN Environment Program: Nature-based solutions*

Non-revenue water: Non-revenue water represents water that has been produced and is "lost" before it reaches the customer (either through leaks, theft, or legal usage for which no payment is made). Reducing non-revenue water can help to increase utility efficiency and allow more funds to be made available for maintenance and further investment, as well as reduce the strain on scarce water resources. *Source: UN Water GLAAS: Financing Water, Sanitation, and Hygiene Under the Sustainable Development Goals*

Open defecation: The disposal of human feces in fields, forests, bushes, open bodies of water, beaches, and other open spaces or with solid waste. *Source: WHO/UNICEF JMP: Sanitation*

Operations and maintenance arrangements: Includes activities necessary to keep services running. Operating costs are recurrent (regular, ongoing) spending to provide WASH goods and services: labor, fuel, chemicals, materials, and purchases of any bulk water. Basic maintenance costs are the routine expenditures needed to keep systems running at design performance, but does not include major repairs or renewals. *Source: UN Water GLAAS: Financing Water, Sanitation, and Hygiene Under the Sustainable Development Goals*

Public sanitation facilities: Facilities that are available to the general public, sometimes for a fee. *Source: USAID Water and Development Strategy Implementation Brief on Sanitation*

Resilience: The ability of people, households, communities, systems, and countries to reduce, mitigate, adapt to, and recover from shocks and stresses in a manner that reduces chronic vulnerability and facilitates inclusive growth. *Source: USAID Building Resilience to Recurrent Crisis Report*

Revolving fund: A mechanism intended to leverage public, private, and donor funding to finance water and sanitation infrastructure. *Source: Millennium Challenge Corporation: Beyond the Impact*

Safe drinking water: Also known as potable water, safe drinking water is considered acceptable for drinking or to use in food preparation. *Source: WHO/UNICEF JMP: Drinking Water*

Safely managed drinking water: Drinking water from an improved water source that is located on-premises, available when needed, and free from fecal and priority chemical contamination. *Source: WHO/UNICEF JMP: Drinking Water*

Safely managed sanitation: The use of improved facilities that are not shared with other households and where excreta are safely disposed *in situ* or transported and treated off-site. *Source: WHO/UNICEF JMP: Sanitation*

Shocks: An acute, short- to medium-term episode or event that has substantial, negative effects on people's current state of well-being, level of assets, livelihoods, or their abilities to withstand future shocks. A shock's onset may be slow or rapid and may affect select households (idiosyncratic shocks) or a large number or class of households (covariate shocks) at the same time. *Source: USAID Global Food Security Strategy 2022-2026*

Social and Behavioral Change: An expansive approach to behavior change that aims to affect key behaviors and social norms by addressing their individual, social, and structural determinants. *Source: USAID Water and Development Technical Series: Social and Behavioral Change for Water Security, Sanitation, and Hygiene*

Stressors: A longer-term pressure that undermines current or future vulnerability and well-being, including—but not limited to—climate variability and change, population pressure, and environmental degradation. *Source: USAID Global Food Security Strategy 2022-2026*

Surface water: Water that comes from rivers, streams, creeks, lakes, and reservoirs. Surface water is also the lowest rung on the JMP drinking water service ladder and is defined as: *drinking water directly from a river, dam, lake, pond, stream, canal, or irrigation channel.* Drinking water from such sources poses the greatest risks to health because of the high risk of contamination. *Source: WHO/UNICEF JMP: Drinking Water*

Sustainability: The ability of a local system to produce desired outcomes over time. *Source: USAID: Local Systems: A Framework for Supporting Sustained Development*

System: A set of interconnected actors whose collective actions produce a particular development outcome. Systems thinking comprises a set of analytic approaches and associated tools that seek to understand how systems behave, interact with their environment, and influence each other. *Source: USAID: Local Systems: A Framework for Supporting Sustained Development*

Underserved: All the individuals, households, or population groups who do not have access to basic services or better.

Unimproved drinking water: Drinking water that comes from an unprotected dug well or unprotected spring. Such sources are difficult to protect from contamination. *Source: WHO/UNICEF JMP*

Unimproved sanitation: The use of pit latrines without a slab or platform, hanging latrines, or bucket latrines. Such facilities enable fixed-point defecation, but do not protect from contact with feces, limiting health benefits. *Source: WHO/UNICEF JMP: Sanitation*

Value chain: The set of actors and activities required to bring products from production to consumption, including processing, storage, transportation, marketing, distribution, and retail. As a product moves through a value chain, each step adds monetary value to the product. *Source: USAID Global Food Security Strategy 2022-2026*

Vulnerability: The propensity or predisposition to be adversely affected. Vulnerability encompasses a variety of concepts and elements including sensitivity or susceptibility to harm and lack of capacity to cope and adapt. *Source: USAID Climate Strategy 2022-2030*

Water quality: Refers to the chemical, physical, biological, and radiological characteristics of water. It is a measure of the condition of water relative to the requirements of one or more biotic species and or human need or purpose. It is most frequently used by reference to a set of standards against which compliance, generally achieved through treatment of the water, can be assessed. *Source: U.S. Geological Survey: Water Quality*

Water resources management (WRM): The process of planning, developing, and managing water resources, in terms of water quantity and quality, within and across water uses for the benefit of humans and ecosystems. WRM includes the institutions, infrastructure, incentives, and information systems that support and guide water management and uses. *Source: USAID Water and Development Technical Series: Water Resources Management*

Water scarcity: Lack of adequate quantities of water for human and environmental uses. While many definitions of water scarcity exist, it is generally considered to be a physical characteristic of the environment and is often quantified in terms of the total water resources available to the population in a given region or country. *Source: UN Water: Water Scarcity*

Water security: The capacity of a population to safeguard sustainable access to adequate quantities of and acceptable quality water for sustaining livelihoods, human well-being, and socio-economic development, for ensuring protection against water-borne pollution and water-related disasters, and for preserving ecosystems in a climate of peace and political stability. Having "water security" implies access to safe drinking water and sanitation services as well as water for agriculture, energy, and other economic activities. *Source: White House Action Plan on Global Water Security*

Water stress: The ability, or lack thereof, to meet human and ecological demands for water. Compared to scarcity, water stress is a more inclusive and broader concept. It considers several physical aspects related to water resources, including water scarcity, but also water quality, environmental flows, and the accessibility of water. *Source: UN Water Report on Progress on Level of Water Stress 2021*

Centers for Disease Control and Prevention Plan

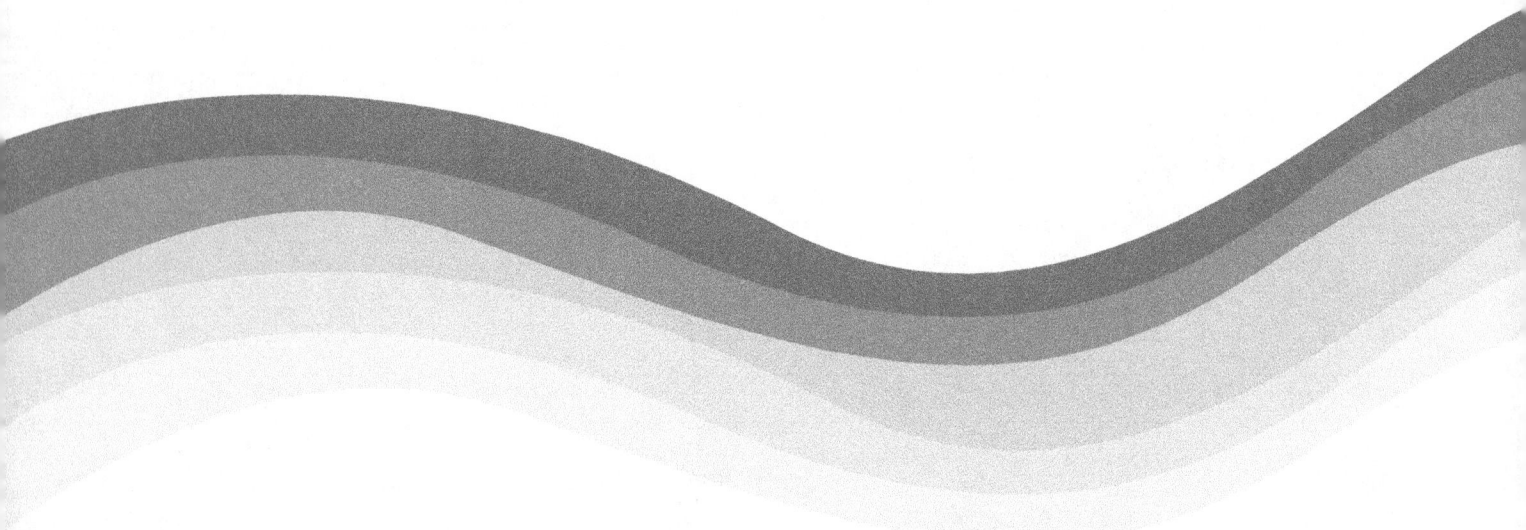

Centers for Disease Control and Prevention

Introduction

The Centers for Disease Control and Prevention (CDC) works 24/7 to protect America from health, safety, and security threats in the United States and other countries. Whether diseases start at home or abroad, are chronic or acute, treatable or preventable, caused by human error or deliberate attack, CDC detects and responds to these disease threats and supports communities and individuals to do the same. CDC increases the health security of our nation and the world. As a public health agency, CDC saves lives and protects people by preventing and detecting health threats, including those related to water. To accomplish our mission, CDC focuses on science. This evidence base provides critical health information that protects our nation and the world against dangerous and costly health threats.

CDC works globally to prevent, detect, and respond to waterborne diseases, such as cholera, hepatitis A and E, and typhoid. CDC also focuses on water, sanitation, and hygiene (WASH)-related diseases and emerging, public health threats, such as malaria, soil-transmitted helminths, antimicrobial resistance, and COVID-19 through multi-disciplinary approaches. CDC has recognized strengths in disease surveillance, laboratory methods and analysis, and monitoring and evaluation of WASH interventions and programs. CDC uses this expertise to identify the most effective strategies in diverse situations ranging from emergency response to longer-term engagement on country, regional, and global WASH and disease prevention and control programs. Throughout this process, the end goals are the same: reducing the impacts of waterborne and WASH-related disease threats through prevention and detection, while strengthening global capacity to prevent and respond to these health risks.

Contribution to the Global Water Strategy

As a global public health agency, CDC believes the success of the U.S. Global Water Strategy (GWS) will require dedicated detection and response capacity to identify existing and emerging waterborne and WASH-related threats. CDC will accomplish this by implementing the activities below, including increasing surveillance for health threats in water, increasing access to and use of WASH services and infrastructure, especially amongst the most vulnerable, and building the institutional capacity of partner agencies to ensure resilient WASH systems. CDC plays an important role domestically and internationally by working with governments and our UN and nongovernmental organization (NGO) partners to detect and respond to new and emerging health threats, strengthen the evidence base to increase access to, and use of, WASH services and infrastructure, and strengthen workforce development and support global public health networks. The agency has decades of on-the-ground epidemiologic, behavioral, environmental, and laboratory experience and programmatic expertise to contribute to WASH programs and advance the GWS.

The goals of the GWS align with CDC's overall strategic priorities. Stronger WASH-related institutions within partner governments will result in an increased capacity to prevent, detect and respond to waterborne and WASH-related health threats in-situ before they emerge into global health threats. Sustainable access to climate-resilient and safely managed drinking water and sanitation services and improved hygiene practices will reduce the risk of disease transmission

1

— including both waterborne and WASH-related diseases, especially those preventable by WASH practices such as hand hygiene — by mitigating many of the most common transmission pathways. Climate change is likely to lead to greater spread of waterborne diseases such as cholera, typhoid, and hepatitis E, especially in vulnerable populations (e.g., low-income communities, those in fragile states, and those living in humanitarian settings).[xliv] Climate change will likely affect the transmission intensity and expand the geographic range of certain WASH-related diseases like malaria, soil-transmitted helminths, and others, and contribute to the emergence and spread of antimicrobial resistance.[xlv,xlvi,xlvii,xlviii] Therefore, addressing climate-related shocks and stressors will be an important aspect of a comprehensive strategy for identifying and mitigating public health vulnerabilities to climate change. Finally, reducing the water-related drivers of conflict and instability will create opportunities to improve the public health of populations in fragile states and other vulnerable situations,[xlix] many of whom face the greatest barriers to improved health outcomes.

CDC's contribution to the GWS is well integrated into the agency's priorities and initiatives. CDC is a principal contributor to the Global Health Security Agenda and its aim to prevent, detect, and respond to global health threats. CDC is currently developing its Climate and Health Strategy and CDC's global WASH activities will be integrated into this strategy. In addition, CDC has decades of experience working in the humanitarian sector and will continue to support efforts to improve the public health of those affected by natural disasters, conflict, and climate-induced threats. CDC also supports the Global Task Force on Cholera Control's Ending Cholera: Roadmap to 2030 initiative. CDC will continue to work at the global, regional, and country levels to strengthen national cholera control and elimination plans and their WASH, epidemiologic, laboratory, and immunization-related components. CDC has also worked for decades collaboratively with foreign ministries of health to strengthen surveillance and laboratory capacity to detect and confirm diseases of public health importance.

CDC's WASH efforts will include a strong focus on emerging WASH-related disease threats, such as COVID-19 and antibiotic resistance. Through the U.S. National Action Plan to Combat Antibiotic-Resistant Bacteria (CARB), CDC leads the U.S. public health response to combat antibiotic resistance, a threat that can continuously emerge and spread across the world. CDC's goals and activities related to detection of existing and emerging antibiotic resistance threats, improving infection prevention and control, evaluating the impact of antibiotic resistance in surface water and wastewater, and evaluating the impact of WASH on antibiotic resistance are also covered explicitly as goals in the CARB National Action Plan. In line with these efforts, CDC supports global initiatives to strengthen WASH and infection prevention and control in healthcare facilities and other institutions and will expand these efforts in the coming years. Linking WASH to such larger global public health initiatives will help raise the profile of WASH within the public health arena.

Results Framework

To effectively realize its contributions to the GWS, CDC will utilize an approach based on the following four pillars.

1. **Strengthen preparedness, detection, response, and recovery**

   CDC will work with partners including U.S. agencies, national governments, and their ministries, UN agencies, and non-governmental partners to provide technical, logistical, and funding support for WASH activities to establish and improve public health preparedness and response capacity; detect and disseminate information about waterborne and WASH-related disease threats; and inform WASH interventions in outbreak, emergency, and post-emergency settings.

2. Collaboration, innovation, and technical guidance

CDC will coordinate and participate in technical working groups, advisory committees, and collaborative partnerships to support global waterborne disease prevention and WASH initiatives by contributing to the establishment of global standards including working closely with the World Health Organization (WHO) to create global guidance. CDC will also inform, develop, revise, disseminate, and update technical guidance and standards related to WASH.

3. Training and capacity building

CDC will develop, pilot, implement, evaluate, and improve trainings, fellowships, and other professional and workforce development activities to strengthen the technical capacity of local and international partners on epidemiology, surveillance, laboratory methods, and monitoring and evaluation. These results of these efforts will prevent disease and guide WASH programming, improving the quality of WASH interventions.

4. Research and Knowledge Management

CDC will plan, conduct, disseminate, and provide technical assistance and review research conducted by CDC and partners to study the etiology of specific waterborne diseases and WASH-related health threats; determine the effectiveness, quality, feasibility, efficiency, and operation of WASH programs, projects, systems, tools, and methods; and develop and/or pilot novel tools, methods, and approaches. CDC will drive the development of the evidence base to improve access to and use of WASH systems.

Priority Areas

CDC will utilize an approach, based on the four pillars above, to carry out activities in five priority areas that are directly related to the objectives of the GWS. These priority areas as well as the GWS strategic objectives (SOs) to which they contribute are described in the following section.

Disease Outbreaks (SO 1 and SO 2)

Preventing, detecting, and responding to waterborne diseases such as cholera, typhoid, and hepatitis as part of the Global Health Security Agenda. This includes supporting the Global Task Force on Cholera Control (GTFCC) *Ending Cholera: Roadmap to 2030* initiative in which water, sanitation, and hygiene play an essential role.

- CDC will continue to support the GTFCC and its technical working groups as well as regional and country-level cholera prevention and response activities.
- CDC will implement integrated WASH surveillance and cholera-related vaccine activities in at least three GTFCC priority countries and support WASH-specific activities in several others. Further expansion will be based on available resources.

- CDC will continue to support typhoid surveillance in the Eastern Mediterranean region, where extensively drug-resistant typhoid fever emerged in 2016, to inform regional and global typhoid prevention and control.

- CDC will continue to develop, pilot, and roll out the Burden and Risk Assessment of Typhoid (BRAT) tool. This will be conducted in at least five countries over the next three years.

- CDC will continue to collaborate with its UN and NGO partners to detect and respond to outbreaks of Hepatitis E, especially among vulnerable and displaced populations.

This work will directly contribute to the strengthening of water and sanitation sector institutions (SO 1) and will indirectly support the conditions for increased access to climate-resilient drinking water and sanitation services and hygiene practices (SO 2).

Strengthening WASH in Healthcare Facilities (SO 1 and SO 2)

Improving access to effective short- and long-term WASH interventions and services in healthcare facilities. Within this priority area, CDC will continue ongoing WASH in healthcare facilities work in 12 to 15 countries in Africa, Latin America and the Caribbean, and the Middle East:

- CDC and its partners will continue to refine WHO's Water and Sanitation for Health Facility Improvement Tool (WASH-FIT), which is a facility improvement tool used to assess WASH services in healthcare facilities and make recommendations on the necessary actions to upgrade services and mitigate risks.

- CDC and its partners will continue to provide technical support to expand the sustainable production and distribution of alcohol-based hand rub, including for healthcare facilities and in outbreak and emergency response scenarios.

- CDC and its partners will engage in evaluations of novel approaches to improving the access to and use of WASH facilities in healthcare settings.

- CDC will coordinate and collaborate with ongoing work on infection prevention and control in healthcare facilities to ensure maximum impact of WASH interventions.

This work will directly contribute to the strengthening of water and sanitation sector institutions (SO 1) and to improving access to drinking water and sanitation services and hygiene practices (SO 2).

Mitigating COVID-19 and Other Emerging Threats (SO 1 and SO 4)

Strengthening WASH mitigation efforts to reduce transmission of COVID-19 and other emerging disease threats, with a focus on identifying, evaluating, and scaling up effective hand hygiene interventions in emergency and non-emergency settings.

- CDC will continue ongoing WASH mitigation efforts focused on healthcare facilities and other community institutions such as schools in up to 15 countries in Central America and the Caribbean, sub-Saharan Africa, and the Middle East. All of these programs focus on low-income populations and more than half target displaced populations, refugees, and those living in urban informal settlements. Continuation of these activities beyond FY 23 will be based on availability of funds.

- CDC will leverage lessons learned in COVID-19 mitigation efforts to inform emerging best practices guidance for WASH in community institutions and public places that have not traditionally been included, as well as behavior change communication for WASH.

Assessing emerging health threats, including antibiotic resistance and COVID-19, in water and wastewater systems within community settings, as well as assessing threats within the broader environment to describe community burdens and ultimately improve prevention, detection and response internationally, through activities outlined in the CARB National Action Plan.

- CDC's Global Antimicrobial Resistance Lab and Response Network improves the detection of existing and emerging antimicrobial resistance threats and identifies risk factors that drive the emergence and spread of resistance across health care, the community, and the environment, including water.

- CDC and its partners will work to strengthen understanding of the effectiveness of WASH systems, especially sanitation systems, at mitigating the transmission of antibiotic resistance in community settings.

- CDC and its partners will work to strengthen technical capacity for and appropriate use of environmental sampling — including wastewater and environmental waters — as supplemental surveillance systems to detect and monitor emerging or re-emerging threats, including monitoring for antibiotic-resistant organisms in low resource settings.

These activities will result in stronger WASH-related institutions (SO 1) and concomitant response to public health threats (SO 4).

WASH in Humanitarian Response (SO 1, SO 2, and SO 4)

Addressing WASH and WASH-related disease threats following natural disasters, conflict, displacement, or among other populations in humanitarian settings.

- CDC will collaborate with U.S. government and UN partners to provide technical assistance to humanitarian responses during the acute emergency phase through the recovery phase via field deployments and technical assistance. CDC has decades of experience on WASH issues in humanitarian and emergency settings, work that directly contributes to reducing instability and fragility, and is therefore directly aligned with GWS SO 4.

- CDC will continue to support Early Warning Disease Surveillance systems and strengthen the response to WASH-related disease threats such as cholera, hepatitis E, and typhoid to mitigate risks and reduce disease-related morbidity and mortality.

- CDC will also support WASH assessments, surveys, and monitoring and evaluation of WASH programs to improve the evidence base in the humanitarian WASH sector. CDC support will be global and based on humanitarian need and U.S. government access.

This work will directly contribute to the strengthening of water and sanitation sector institutions (SO 1), increase access to WASH services (SO 2), and mitigate the effects of water-related conflict, instability, and fragility (SO 4).

Waterborne Disease, WASH, and Climate Change (SO 2 and SO 4)

Addressing public health threats of climate change on waterborne disease transmission as well as the impacts of extreme weather events on vulnerable populations.

Climate change will impact the availability of freshwater resources, degrade water quality in many areas and contribute to overall water stress for hundreds of millions of people globally. Extreme weather events and increased flooding will degrade water quality and displace populations, further exacerbating existing vulnerabilities.[i] CDC is implementing the following activities to address climate-related threats:

- Develop and pilot early warning systems for cholera and other waterborne diseases in areas vulnerable to flooding and identify climate parameters linked to cholera outbreaks to further refine modeling and forecasting for cholera in cholera hotspots.

- Develop and pilot water quality monitoring programs for locations impacted by drought, saltwater intrusion, and other impacts of climate change.

Principles

CDC will apply the principles of the GWS by leveraging the agency's unique capacities of scientific leadership and expertise, public health workforce development, laboratory capacity and innovation, and data analytics to drive impact. CDC supports a whole-of-government approach and will continue to work with all levels of government, both domestic and international, using public health data to inform decisions and establish systems for emergency operations and response to strengthen global health security. CDC is committed to health equity in its programming and continues to focus on meeting the needs of the most vulnerable and underserved people and communities. CDC understands the importance of the trust placed in the science and technical leadership of the agency to keep Americans safe from public health threats both domestically and abroad, and therefore is committed to following scientific best practices in research, learning, and innovation. CDC also recognizes the urgency of integrating climate change adaptation and resilience into all aspects of programming.

Resource Implications

Pending available resources, CDC will continue to support water and global WASH activities by contributing dedicated technical and programmatic staff to provide technical assistance and programmatic support, as well as financial resources that will be used to procure materials and fund cooperative agreements with its partners. CDC will continue to leverage funding for Global Health Initiatives such as the Global Health Security Agenda and COVID-19-related funds to support cross-cutting WASH activities for disease prevention. CDC will continue to support work outlined in the CARB National Action Plan related to this work, however, more resources are needed to meet the goals outlined in this plan.

Assumptions

CDC's contributions to the GWS will be contingent on available funding for the relevant activities and programs described above. The activities described above are based on similar funding levels over the next five years.

INTERNATIONAL
TRADE
ADMINISTRATION

International Trade
Administration Plan

International Trade Administration

Introduction

The International Trade Administration's (ITA) mission is to strengthen the competitiveness of U.S. industry by promoting trade and investment, ensuring fair trade, and enforcing trade laws and agreements. ITA staff are located in more than 100 U.S. cities and 75 global markets and work directly with U.S. firms in the water sector to help individual companies export their goods and services overseas. At ITA's headquarters in Washington, DC, industry analysts monitor the competitiveness of the industry both at home and abroad. Through collaboration across the organization, ITA gathers valuable information about opportunities and trade barriers that exist in various foreign markets and disseminates it to public and private sector stakeholders. In addition, ITA staff employ their expertise in support of U.S. water and wastewater treatment companies at trade promotion events, including trade missions and trade shows, and through a variety of customized in-country services. ITA officials regularly advocate on behalf of U.S. companies competing for public water and wastewater tenders overseas. They engage in commercial diplomacy with foreign governments to help resolve individual commercial disputes through staff-level meetings and address industry-wide trade barriers through high-level bilateral and multilateral dialogues.

Contributions to the Global Water Strategy

Ongoing ITA activities contribute to the U.S. Global Water Strategy's (GWS) goal of building health, prosperity, stability, and resilience through sustainable and equitable water resources management, access to safe drinking water and sanitation services, and hygiene practices. As part of its mission to strengthen the competitiveness of U.S. industry by promoting trade, ITA works with companies in the water and wastewater treatment sectors to help them deploy technologies, goods, and services to markets around the globe. Examples of water solutions provided by U.S. companies include municipal drinking water treatment and delivery, municipal wastewater conveyance and treatment, ground and surface water remediation, and industrial water treatment solutions. The sector also provides analytical, consulting, and engineering services for water projects.

Strategic Objective 1: Strengthen Water and Sanitation Sector Governance, Financing, Institutions, and Markets

ITA contributes to the GWS SO 1 through the following activities:

Resources for U.S. companies

ITA produces expert industry analysis that enables U.S. companies to better identify and understand opportunities in emerging markets for their products and services, thereby growing trade and strengthening markets for innovative water technologies.

- **Top Export Market Rankings:** ITA publishes an annual *Environmental Technologies Top Export Market Rankings* report that helps companies determine their next export market by comparing opportunities across borders. The report includes case studies for key markets where the scope of opportunity for U.S. companies is limited by trade or other barriers, and it highlights which technologies and services in the water sector are in demand in the case study markets. The report also informs the U.S. government about where policy interventions could enable better environments for doing business abroad.

- **Other market intelligence:** ITA provides additional market intelligence such as the *Environmental Technologies Resource Guide,* which consists of country snapshots featuring information on market opportunities, challenges, key upcoming trends and events, and points of contact. Water sanitation, wastewater treatment, and water management are among the featured market opportunities.

Trade promotion

Through activities such as trade shows and trade missions, ITA connects U.S. companies to foreign buyers and communities in need of water and sanitation solutions.

- **The Water Environment Federation Technical Exhibition and Conference (WEFTEC):** ITA typically provides annual trade event services for WEFTEC, the largest conference for water quality professionals in North America. Services include counseling U.S. companies, bringing foreign delegations to the conference, matching U.S. companies with foreign buyers, and communicating the resources available from ITA. Through a virtual platform, ITA offers appointments for business-to-business and business-to-government meetings. On average, WEFTEC has about 1,000 exhibitors and 22,000 attendees. About 79 international markets are represented.

- **Singapore International Water Week (SIWW):** Led by local embassy staff, ITA supports U.S. exporters participating in SIWW, a biannual event. In addition to networking opportunities and market counseling, ITA works with the U.S. pavilion organizer to maximize the exposure and benefit to those exhibiting companies.

- **Customized counseling and other services:** ITA regularly engages with U.S. companies in need of export assistance. Examples of successes in the water sector include:

  - In 2022, through customized services such as counseling and an initial market check, ITA helped a leading U.S. provider of wastewater and desalination solutions enter the Ghanaian market for the first time through a local partner who can help offer distributed water solutions.

  - After working with ITA staff for two years, a U.S. provider of water metering solutions signed an agreement with a distributor in Kenya and made its first sale to that market in 2021.

  - ITA recently helped a U.S. firm secure a contract funded by the Asian Development Bank to provide technology solutions for a climate adaptation project designed to protect coastal districts in India from flooding.

- **U.S. Water Partnership:** ITA works closely with the U.S. Water Partnership in support of its Water Smart Engagements Program (WiSE) in the Association of Southeast Asian Nations (ASEAN). The program pairs members of the ASEAN Smart Cities Network (ASCN) with U.S. cities, water districts, and utilities to create more innovative, efficient, and effective approaches to 21st-century water challenges. The purpose of WiSE is to increase water security in ASEAN partner cities through sustainable water management solutions; to establish long-term relationships between ASEAN and U.S. utilities to foster communication and build capacity; and to increase the exchange of services, goods, science, and technologies. ITA supports WiSE activities at WEFTEC and at SIWW.

Diplomatic engagement

ITA promotes policies in foreign markets to enable trade and investment in products and services in the water sector, which facilitates greater market access for U.S. companies and helps ensure a more conducive climate for local business as well.

ITA staff regularly engage with foreign governments to conduct commercial diplomacy and work with U.S. companies to help them enter or expand their presence in these markets. Examples of commercial diplomacy objectives include reducing or eliminating tariffs, advocating for a transparent and predictable regulatory climate, supporting intellectual property rights protection and enforcement, and seeking the adoption and implementation of strong, industry-led environmental standards.

Global Water Strategy Principles

ITA's contributions to the GWS are informed by the Department of Commerce's Strategic Plan for 2022-2026, which is highly compatible with the GWS principles. Three concepts drive the Commerce Department's work: innovation, equity, and resilience. The Commerce Department's SOs include: 1) drive U.S. innovation and global competitiveness; 2) foster inclusive capitalism and equitable economic growth; 3) address the climate crisis through mitigation, adaptation, and resilience efforts; 4) expand opportunity and discovery through data; and 5) provide 21st-century service with 21st-century capabilities.

To illustrate the alignment between the Commerce Department's Strategic Plan and the GWS principles, the principle "Focus on meeting the needs of marginalized and underserved people and communities and those in vulnerable situations" is in sync with the Commerce Department's SO 2, which strives to create an economy that works for all Americans. Specifically, as of 2022, ITA has planned an executive-led trade mission to Italy, Spain, and Portugal aimed at minority-owned businesses as well as another executive-led trade mission to France, Netherlands, and Portugal for women in technology. Another GWS principle, "Leverage data, research, learning, and innovation," is reflected in the Department's SO 1, SO 3, and SO 4. ITA's industry analysts are committed to using the best tools of data and analysis to support the export of innovative environmental technologies. Finally, resilience is a core theme to both the GWS and to the Department of Commerce, as indicated in SO 3 in particular.

Geographic Focus

ITA's focus is global with support from staff in 75 countries. ITA assists U.S. companies in finding new markets and expanding existing export markets for their equipment, goods, and services. Analysts at ITA work to identify those markets that hold the most promise for expanding U.S. exports and are where U.S. government resources can be most effectively deployed. Export markets for the water industry include countries seeking to improve access to safe drinking water and sanitation services and to manage water resources more equitably and sustainably.

Resource Implications

ITA's contributions to the GWS, as described in this plan, represent core functions of the agency and are therefore core budget items with significant allocations of agency funds and staff time.

Assumptions

The ITA contributions outlined in this plan assume three things: that there is at least a flatline of funding to support existing activities that support U.S. water and wastewater treatment companies; that full staffing requirements are met; and that in the period 2022-2027, no changes occur to the agency's strategic direction or factors that would limit the fulfillment of these contributions.

Millennium Challenge Corporation Plan

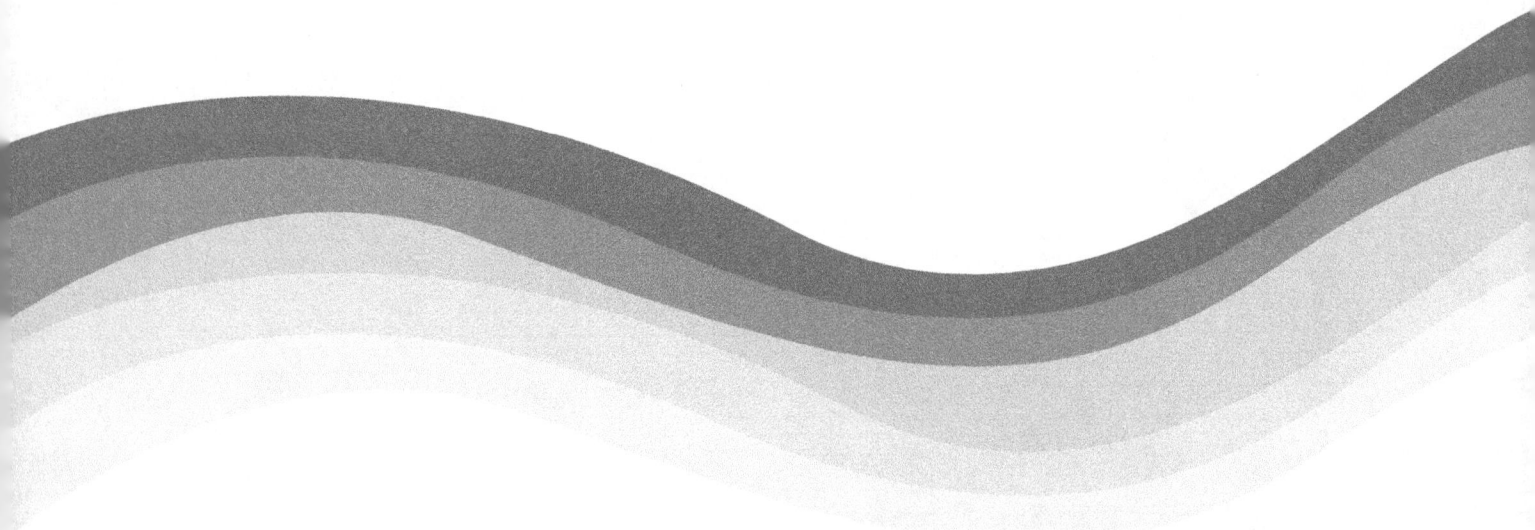

Millennium Challenge Corporation

Introduction

Created by the U.S. Congress in January 2004 with strong bipartisan support, Millennium Challenge Corporation (MCC) has changed the conversation on how best to deliver smart U.S. foreign assistance by focusing on good policies, country ownership, and results. MCC provides time-limited grants promoting economic growth, reducing poverty, and strengthening institutions. These investments not only support stability and prosperity in partner countries but also enhance American interests.

MCC's mission is to reduce poverty through economic growth.

MCC forms partnerships with developing countries that are committed to good governance, economic freedom, and investing in their citizens. MCC is a prime example of smart U.S. government assistance in action, benefiting both developing countries and the American taxpayers through:

- Competitive selection

- Country-led solutions

- Country-led implementation

- Focus on results

Contribution to the Global Water Strategy

The MCC offers three types of programs to countries committed to good governance, economic freedom, and investing in their citizens:

- Compacts – large, five-year grants for selected countries that meet MCC's eligibility criteria;

- Concurrent Compacts for Regional Investments – grants that promote cross-border economic integration and increase regional trade and collaboration; and

- Threshold Programs – smaller grants focused on policy and institutional reforms in selected countries that come close to passing MCC's eligibility criteria and show a firm commitment to improving their policy performance.

During the first phase of a grant program development process, MCC and the selected partner country jointly conduct a constraints-to-growth analysis. This analysis identifies the constraints to private investment and entrepreneurship that are most binding to economic growth in the country. The results of this analysis enable the country, together with MCC, to select activities most likely to contribute to inclusive, sustainable, and poverty-reducing growth.

In those countries where a water-related constraint-to-growth is identified, MCC identifies complementary interventions in both hard infrastructure and policy and institutional strengthening as the key element to delivering benefits and promoting sustainability. Accounting for and responding to climate change – both broad trends and variability – is also a key element of assuring that benefit streams are robust under a range of future conditions.

MCC activities traditionally contribute to all four U.S. Global Water Strategy (GWS) SOs. However, it is difficult to make forward projections since future compacts are identified through joint analysis with partner countries. The section below provides examples of how past compacts have contributed to each SO and represent MCC's intentions moving forward.

Results Framework

MCC is committed to delivering inclusive and sustainable economic growth and reducing poverty throughout the entire lifecycle of its investments. MCC's evidence-based approach is rooted in that mission, and its comprehensive results framework seeks to measure, collect, and report on the outputs, outcomes, and impacts of MCC investments. This includes results reporting of policy and institutional reforms associated with initial and ongoing program eligibility, tracking activity inputs and outputs to ensure that projects progress as expected, monitoring interim outcomes as programs near completion, and measuring long-term impacts after program closure through independent evaluations.

MCC's results framework is designed to foster learning and accountability, and it has served as a model for global dialogue about development results and aid effectiveness for over a decade.

Strategic Objective 1: Strengthen Water and Sanitation Sector Governance, Financing, Institutions, and Markets

MCC pairs its infrastructure investments with appropriate interventions to strengthen utility and sector governance and, where possible, seeks to engage private sector participation in financing and operations of that infrastructure. The following provides examples of MCC's recent work in this area:

- **Jordan – Public-Private Partnership**: The objectives of the As-Samra Expansion Project were to (i) increase the capacity to treat wastewater from Amman and Zarqa Governorates, (ii) increase the volume of treated wastewater that is available as a substitute for freshwater for non-domestic use, and (iii) release the freshwater used by the agricultural sector for municipal use. Under a project finance public-private partnership, with help and a funding commitment from MCC, the plant underwent an expansion and technological upgrades, which allowed the government to treat 70 percent of the country's wastewater and meet the region's wastewater treatment needs through 2025. The expanded plant provides 133 million cubic meters of high-quality treated water per year – equivalent to over 10 percent of Jordan's entire annual water resources – for irrigation in the Jordan Valley.

- **Zambia – Master Planning**: Investment master plans for both water supply and sanitation were drawn up during the development of the Zambia Compact, and a stormwater management master plan was developed during compact implementation. These master plans not only helped the Government of Zambia prioritize MCC investments, but also added significant value beyond the compact investment by outlining clearly to the private sector and donor community where funding was necessary, resulting in hundreds of millions of dollars of leveraged funding.

- **Cabo Verde – Creating a Regionalized Water Company**: The objective of the Utility Reform Activity was to improve the delivery of water and sanitation services on the main island of Santiago through the creation of an island-wide, publicly owned water and sewer company operating on standard commercial principles. Small and fragmented water departments were unable to operate at scale and provide high-quality service to customers. Moving to a single, corporatized, island-wide utility was expected to achieve the economies of scale necessary to professionalize service delivery and increase efficiency, effectiveness, and quality of service. As a result of the compact, the number of service delivery hours per day has increased by nearly two hours and residential water consumption has increased by over seven liters per capita per day.

Strategic Objective 2: Increase Equitable Access to Safe, Sustainable, and Climate-Resilient Drinking Water and Sanitation Services and the Adoption of Hygiene Practices

- **Mongolia – Water Supply**: MCC's investment aims to provide a sustainable supply of water that will stem the impact of an impending water crisis and sustain private sector-led economic growth in the capital city, Ulaanbaatar. The MCC compact will increase the available supply of water through infrastructure for the development of new wellfields and capacity building. The compact's Water Supply Project will support the construction of new groundwater wells and a state-of-the-art plant for purifying drinking water, the construction of a new plant for treating wastewater, and the development and implementation of policy, legal, regulatory, and institutional reforms that enhance the long-term sustainability of Ulaanbaatar's water supply. These investments will increase the supply of water to Ulaanbaatar by more than 80 percent, putting the city on a better footing to sustain its future economic growth.

- **Timor Leste – Water Supply and Sanitation**: The planned compact is focused on a Water, Sanitation, and Drainage Infrastructure Project that aims to reduce widely prevalent fecal pathogen contamination in water resources and the environment. This contamination impedes effective improvements in both child and adult well-being due to frequent and extended incidences of diseases. The proposed project includes constructing a plant to produce water disinfectant for city water supply, building the country's first central wastewater system, improving the drainage network for the capital city of Dili, policy and institutional reform, and social and behavior change. The project will ultimately supply disinfected water to 429,000 residents in Dili and 64,000 residents in the surrounding area.

Strategic Objective 3: Improve Climate-Resilient Conservation and Management of Freshwater Resources and Associated Ecosystems

- **Niger – Surface Water Irrigation**: In Niger, the agricultural sector employs more than 80 percent of the population and represents the second-largest export sector. However, due to frequent droughts and floods that decimate crops and productive assets, much of the Nigerien population struggles to maintain a subsistence existence, let alone increase household incomes. Although poverty trends are slowly improving, with a climate prone to volatile weather conditions and a lack of access to critical inputs and information, agricultural productivity has stagnated. The compact includes investments in irrigation infrastructure and management systems, climate-resilient agricultural production, upgraded roads to improve market access, and management of natural resources, while seeking to empower entrepreneurs and smallholder producers.

- **Lesotho – Surface Water Irrigation**: The Irrigation and Water Reform sub-activity will strengthen the capacity of key government ministries responsible for providing water and irrigation-related services necessary for MCC-targeted investments, in particular the Ministry of Water, the office of the Commissioner of Water, and the Ministry of Agriculture, Marketing and Food Security. It will support the creation of an appropriate legal framework for irrigated agriculture, including the creation of water user associations. It will support revisions to the Water Act in order: (i) to promote improved water resources management (WRM) and support the implementation of key WRM activities; and (ii) to strengthen the legal basis for Integrated Catchment Management. The sub-activity will also support cross-cutting measures to improve collaboration between relevant government agencies at the national and local levels to foster replication of the project's success. Inter alia the project aims to reduce the economic and climate vulnerability of farmers.

Strategic Objective 4: Anticipate and Reduce Conflict and Fragility Related to Water

- **Niger – Adaptation to Climate Change**: The Climate-Resilient Communities (CRC) Project aims to increase incomes for small-scale agriculture- and livestock-dependent families in eligible municipalities in rural Niger. Through two activities, the Regional Sahel Pastoralism Support (PRAPS) Activity and the Climate-Resilient Agriculture Activity, the compact will support the recovery and restoration of public land pastures, sustainable increases in farm productivity, and income generation. They seek to strengthen the resiliency of farmers and agro-pastoralists to climate shocks through an integrated climate-resilient investment plan. The project aims to support the restoration and conservation of about 5,400 hectares of pasture lands, about 55,000 hectares of agriculture land, and up to 530 hectares of small-scale irrigation.

Principles

The key principles that guide MCC's work include:

- Country-led program design and implementation which requires working through local, regional, and national systems
- Singular focus on poverty reduction and inclusive growth
- Evaluating all our investments and drawing lessons learned for our successes and failures; and
- Building in resilience and sustainability to ensure that benefits continue to accrue over a range of conditions.

By statute, MCC has close ties to USAID – both at the board level and at an operational level. This interagency collaboration includes coordination of program activities and USAID's role in helping prepare candidate countries for eligibility by improving their performance against MCC's scorecard.

MCC is committed to delivering inclusive and sustainable economic growth and reducing poverty throughout the entire lifecycle of its investments. MCC's evidence-based approach is rooted in that mission and its comprehensive results framework seeks to measure, collect, and report on the outputs, outcomes, and impacts of MCC investments. MCC's commitment to making decisions based on data and evidence pervades all stages of the agency's engagement in a country. MCC country scorecards, constraints analyses, cost-benefit analyses, and monitoring and evaluation plans are some of the main tools MCC uses to achieve, measure, learn from, and transparently report its results. This evidence informs and shapes future project design and decisions.

Climate change poses the greatest risk to developing countries whose people, economies, and institutions are less able to adapt or afford its consequences, and who bear the largest burden of those impacts. Climate change directly affects MCC's mission to reduce poverty through inclusive and sustainable economic growth. Without ambitious and targeted interventions, climate change will reverse significant development gains made in these countries and exacerbate global poverty and inequality. To elevate our climate ambition, MCC will promote low-carbon economic development to support countries' transitions away from fossil fuels, maintain a coal-free policy across our investment portfolio, and commit more than 50 percent of our program funding to climate-related investments – such as projects that are resilient to climate change, help countries adapt, and help reduce greenhouse gas (GHG) emissions.

MCC seeks to fund activities that will generate significant and measurable increases in the incomes of large numbers of people in partner countries, including significant gains for the poor. As part of MCC's commitment to inclusion, MCC designs investments to ensure benefits for women, the poor, and other disadvantaged groups. Each MCC investment requires a Social and Gender Integration Plan, which provides a comprehensive roadmap for social inclusion and gender integration throughout compact and threshold programs.

Geographic Focus

Each year, the MCC Board of Directors selects countries as eligible for MCC assistance. A hallmark of MCC's model is transparency about the process and criteria that govern the selection of country partners, including:

1. Low- or lower-middle-income countries (LICs/LMICs) are not prohibited from receiving assistance;
2. Updating scorecard indicators;
3. Assessing countries against scorecard; and
4. Selection based on scorecard, opportunities to reduce poverty and inclusive economic growth, and availability of funds.

Consequently, MCC has no geographic (regional) focus areas.

Resource Implications

It is difficult for MCC to make forward projections of financial contributions since its partner countries and their constraints to economic growth at a particular point in time are difficult to forecast. MCC activities traditionally contribute to all four SOs. However, it is difficult to make investment projections in the Water Sector since future compacts are identified through MCC's competitive selection and constraints analysis processes are used to identify the priority sector(s) in each of our partner countries. The section below provides examples of how past compacts have contributed to each SO and represent MCC's intentions moving forward.

Assumptions

MCC's planned contributions to the GWS may be disrupted by suspension or termination of any of our relevant Compacts. MCC recognizes that a compact-eligible country can generally maintain or even improve policy performance, but not meet the formal eligibility criteria each year due to several reasons. If a country does demonstrate a significant policy reversal, MCC may issue a warning, suspend, or terminate eligibility and/or assistance.

Even after a country has been selected for compact eligibility or threshold program assistance, MCC regularly reviews its partner countries' policy performance throughout the development and implementation of a compact or threshold program. As part of this review, MCC may engage in a policy dialogue with partner countries, coordinating with our U.S. Government colleagues at the Department of State, USAID, and U.S. embassies regarding the country's commitment and adherence to the MCC selection criteria.

According to MCC's statute, a country may have its eligibility for assistance suspended or terminated if the country has (i) engaged in activities contrary to the national security interests of the U.S., (ii) engaged in a pattern of actions inconsistent with MCA eligibility criteria, or (iii) failed to adhere to its responsibilities under an MCC program agreement.

The National Aeronautics and Space Administration Plan

National Aeronautics and Space Administration

Introduction

As a leading authority on water research, the National Aeronautics and Space Administration (NASA) seeks to support the U.S. Global Water Strategy (GWS) through its unparalleled capacity to monitor and model the global hydrological system. NASA accomplishes this primarily through basic research, analysis, and applied science programs, with a long legacy of supporting capacity building and decision makers in water-related fields.

The National Aeronautics and Space Act of 1958 established NASA's charter and outlines multiple objectives that align with the aims of the GWS. In particular, it mandates the Agency contribute to:

(1) "The expansion of human knowledge of the Earth…;"

(2) "The preservation of the role of the United States as a leader in aeronautical and space science and technology and in the application thereof to the conduct of peaceful activities within and outside the atmosphere;" and

(3) "The most effective utilization of the scientific and engineering resources of the United States, with close cooperation among all interested agencies of the United States in order to avoid unnecessary duplication of effort, facilities, and equipment"

In service of this mandate, NASA seeks to improve the observation, modeling, and forecasting of water resources through application development, modeling expertise, fundamental domain research, and training. Internally, NASA has already begun developing a unified International Water Strategy with the aim of more efficiently linking the Agency's technical capabilities with global water needs. The results of this framework will improve NASA's ability to mobilize personnel and computing resources and will, in turn, support the aims of the GWS objectives established herein.

Contribution to the Global Water Strategy

NASA exists primarily as a science-based organization that addresses multiple facets of water management through its research programming. The Earth Science Division (ESD) within the Science Mission Directorate (SMD) contains several programs that either directly or indirectly focus on water resources, with laboratories dedicated to the hydrological, biospheric, cryospheric, and ocean ecological sciences. These programs contribute to fundamental Earth systems research using satellite observations, integrated models, and field measurement campaigns. All data generated under these programs are free and open to the public with training and resources available to facilitate access.

ESD's Applied Sciences Program elements such as the Water Resources, Agriculture, Disasters, and Capacity Building address water-related needs – such as flooding, drought, or water quality monitoring – for defined stakeholders and at operational time scales. Regional programs like the Western Water Applications Office (WWAO) target communities in specific geographies with acute water resources needs, while food security initiatives like the NASA Harvest program seek to connect water availability to agricultural and human development. Climate modeling and resilience issues are also core to the Agency capabilities. Dedicated modeling centers like the NASA Center for Climate Simulation and the Global Modeling and Assimilation Office contribute to fundamental understanding of global surface and atmospheric changes.

Interagency cooperation is also critical to NASA's operations, and many of its flagship initiatives such as Landsat (U.S. Geological Survey (USGS)), SERVIR (USAID), and Harvest (University of Maryland) are jointly managed with other U.S. agencies or external partners. The Land Information System (LIS) Hydro system is applied operationally by the US Air Force 557th Weather Wing for improved global hydrologic monitoring and forecasting. In addition, the Group on Earth Observations (GEO) Global Water Sustainability (GEOGLoWS) initiative is an interagency collaboration that promotes the GEO data sharing principles and data management for improved water resource management.

NASA will continue to support these programs, as well as existing water-specific partnerships such as the Interagency Water Working Group (IWWG) and its corresponding applied science arm, the IWWG Science and Applications Team (ISAT). Expanding cooperative programs, such as Strategic Hydrologic and Agricultural Remote-sensing for Environments (SHARE) will help NASA to reach new stakeholders by leveraging connections with the U.S. Department of State and others.

Finally, NASA contributes to understanding water-related socioeconomic issues through our Socioeconomic Data and Applications Center.

Results Framework

Strategic Objective 1: Strengthen Water and Sanitation Sector Governance, Financing, Institutions, and Markets

NASA has several projects in place that help to inform transboundary water governance. For example, SERVIR is helping Bangladesh Water Development Board with streamflow estimates and forecasts, which are not shared by upstream countries. Use of satellite data and modeling tools has enabled the Bangladesh government to take early action ahead of impending floods. Similarly, the sub-seasonal and seasonal forecasts and capacity development activities by SERVIR are helping governments of Nepal, Cambodia, and Vietnam to prepare for droughts when forecasts provide warning signs. Additional value-added products like decision-support systems are used to aid governments in monitoring and forecasting agricultural outputs and water resources — capabilities which also have important implications for SO 4.

Strategic Objective 2: Increase Equitable Access to Safe, Sustainable, and Climate-Resilient Drinking Water and Sanitation Services, and Adoption of Key Hygiene Behaviors

The SERVIR program is also helping to implement satellite-based streamflow forecasting to develop sustainable solutions for safe drinking water in rural communities. One project in Uganda has allowed the local government to develop a water supply system for one water management zone, with plans to expand the methodology to other areas in the future. Leveraging Earth observation methods, approaches, and principles to tackle critical water challenges and improve equity and benefits for disadvantaged, marginalized, and other vulnerable groups is a core component of SERVIR's own strategic goals and this methodology could feasibly be introduced in additional countries on the USAID High-Priority Country (HPC) list.

Strategic Objective 3: Improve Climate-Resilient Management and Conservation of Freshwater Resources and Associated Ecosystems

Sustainable access to climate-resilient supplies of freshwater requires a fundamental understanding of how much water is available and how those resources are likely to change in the future. NASA, through its expertise in Earth-observing systems and integrated modeling, can contribute to the estimation of water quantity. Satellite-based sensors such as the Gravity Recovery and Climate Experiment (GRACE) mission can monitor the distribution of groundwater and reservoirs and identify regions experiencing rapid changes in water storage. Radar and altimetry missions such as ICESat-2 and Sentinel-6 Michael Freilich (S6MF) can measure changes to ice sheet volume, sea levels, and numerous other valuable freshwater indicators. Future missions will further improve the Agency's ability to monitor water resource availability. The future NASA-ISRO Synthetic Aperture Radar (NISAR) satellite will be able to measure changes in groundwater reserves, map soil moisture, and detect surface flooding for a range of hazard applications. The Surface Water Ocean Topography (SWOT) mission, planned to launch in 2022, will provide the first global survey of Earth's surface water, observe the fine details of the ocean's surface topography, and measure how water bodies change over time.

NASA can also support efforts to monitor water quality. The Cyanobacteria Assessment Network (CyAN) is a multi-agency initiative that has developed an early warning system for algal bloom detection. The future Plankton, Aerosol, Cloud, ocean Ecosystem (PACE) mission, planned to launch in 2024, will provide advanced observations of global ocean color, biogeochemistry, and ecology that will be used to identify the extent and duration of phytoplankton blooms and improve understanding of water and air quality. Through coordination with the Water Resources Program and the Global Partnerships Program, NASA is supporting several freshwater studies and applications, including the development of a Freshwater Health Index to guide ecological management of water systems. Many of the aforementioned capabilities are already in use in operational, decision-support contexts, including the use of open-source methods like smartphone applications. Continued support through training could improve stakeholders' ability to access and apply these data for their own planning purposes.

Strategic Objective 4: Anticipate and Reduce Conflict and Fragility Related to Water

NASA supports multiple programs that directly or indirectly monitor water-related drivers of potential societal instability and fragility. Drought and food insecurity are two potential areas in which NASA supports water resource monitoring, through its ability to track these resources across time and space. Operational data products from the ECOsystem Spaceborne Thermal Radiometer Experiment on Space Station (ECOSTRESS), the Famine Early Warning Systems Network (FEWS NET), and the Group on Earth Observations Global Agricultural Monitoring (GEOGLAM) all contribute to decision-support systems for monitoring agriculture and water conditions.

Disaster events are another potential vector for social and ecological instability. Impacts of wildfires, flooding, and tropical storms could exacerbate conditions of resource scarcity and fragility. For scenarios surrounding water stress, food security, and disaster resilience, NASA can provide early warning and post-event response capabilities through its Applied Sciences teams. NASA routinely responds to international disaster charter activations to provide data and imagery in the wake of an event.

Water scarcity and impact on human life are pronounced in West Africa. In the arid regions of Senegal, climate change is impacting the availability of water in traditional watering holes along migratory routes for nomadic tribes. SERVIR is helping keep track of water signatures in those ponds using satellite data and making that information available to nongovernmental organizations (NGOs) on the ground. NGOs relay the information to the pastoralists in search of water for their herd. SERVIR plans to expand on this effort to include short-term and seasonal forecasts to provide a longer-term perspective. This effort is also being replicated in East Africa where similar challenges are present. Together with NASA, USAID is exploring how such efforts provide a stabilizing force for the nomadic population and reduce risks of making fragile populations more vulnerable.

NASA's Commitment to GWS Guiding Principles

NASA embodies several of the GWS Guiding Principles and will continue to uphold these standards with future programs and initiatives. NASA consistently remains at the forefront of scientific research and innovation by adhering to best practices for research, computing, data management, and knowledge dissemination. Climate change adaptation and resilience is of particular importance; NASA translates Earth observation insights about climate change risks to stakeholders across the public and private sectors to address the system causes of risks and foster resilient communities. In addition, NASA recently released its own Climate Action Plan, which addresses NASA's own efforts to increase its own resilience across its field centers. Additionally, NASA maintains strong cooperation with science agencies such as USGS, the Environmental Protection Agency (EPA), and the National Oceanic and Atmospheric Administration (NOAA), as well as intergovernmental research consortiums (e.g., Global

Partnerships Program, GEOGLoWS, and the World Meteorological Organization). These ties serve to help position NASA as a world leader in Earth system research.

Addressing the needs of underserved and marginalized communities is a core mission of NASA Capacity Building programs (i.e., Applied Remote Sensing Training (ARSET), SERVIR, DEVELOP). These, and other initiatives like the Indigenous Peoples Pilot project, provide individuals and institutions with workforce development, training, and collaborative projects to strengthen understanding of Earth observations and expand their use around the world. NASA also aims to strengthen global water access, sanitation, and hygiene (WASH) through its contribution to the White House Action Plan on Water Security. Through these activities, NASA Earth observations can also be used to address gender disparities in water access and quality, increasing our understanding of water resource impacts on women and girls.

In addition to addressing the needs of our partners, stakeholders, and end-users, NASA is committed to answering the needs of its diverse workforce. In response to Executive Order 13583, Establishing a Coordinated Government-wide Initiative to Promote Diversity and Inclusion in the Federal Workforce, NASA has established a dedicated program for DEIA which prioritizes reinforcing a culture in which our employees feel they can be authentic, welcomed, respected, included, and engaged. NASA recently launched its own Equity Action Plan with the aim of further identifying and removing the barriers that limit opportunity in historically underserved and underrepresented communities. In doing so, the Agency will work to enhance grants and cooperative agreements to advance opportunities, access, and representation for underserved communities and to ensure every community benefits from critical Earth science data.

Finally, NASA is committed to open science, with an emphasis on freely available, accessible, and reproducible results. A new data initiative, Transform to Open Science (TOPS) will accelerate the engagement of the scientific community in open science practices through events and activities aimed at: (1) lowering barriers to entry for historically excluded communities; (2) improving understanding of how people can use NASA data and code to help solve issues in their communities; and (3) increasing opportunities for collaboration while promoting scientific innovation, transparency, and reproducibility.

Geographic Focus

Much of NASA's capabilities utilize remote observations that are available operationally and at a global scale. Therefore, the Agency operates largely without geographic limitation on projects it can feasibly support. With that said, NASA also maintains several long-term research programs with specific geographic focuses that will continue to be a large part of its IWWG project portfolio.

The GEOGLoWS, Harvest, and SERVIR programs are some of NASA's flagship internationally-reaching research and capacity-building initiatives, which promote stakeholder engagement through partnerships with multiple regional research facilities. The Eastern & Southern Africa and West Africa SERVIR hubs have previously supported water-related data products and trainings for the following FY 22 HPCs: Ethiopia, Ghana, Kenya, Rwanda, Senegal, South Sudan, Uganda, Zambia, and Tanzania. Additionally, the Amazonia and Himalaya hubs have conducted projects for HPCs Guatemala and Nepal, respectively.

Resource Implications

ESD maintains numerous programs that intersect with the aims of the GWS. Primary research on each aspect of the water cycle is conducted at each center through domain-specific laboratories. Current joint satellite missions with other U.S. and foreign government agencies (e.g., Landsat, MODIS, Sentinel 6B, and IceSat-2) provide a long-term record of global water resources. Future missions such as SWOT and NISAR satellites will further improve our holistic understanding of the global hydrological system by continuing ongoing water observations and providing innovative new data products. At the Agency level, funding for Earth Science is requested to increase 15.3 percent between FY 2023 and FY 2026 which will allow for continued support of these Agency-wide programs.

Additional efforts will be contributed at the program level, with extensive research, training, development resources available through NASA's capacity-building efforts as well as the Applied Sciences Programs for Water Resources, Agriculture, and Disasters. In addition to these internal programs, NASA will continue to engage closely on its interagency work with the IWWG, ISAT, and other partners that support all-of-government water projects.

Assumptions

Much of the work being done in ESD is informed by long-term guidance plans such as the National Academies of Sciences, Engineering, and Medicine's Decadal Survey for Earth Science and Applications from Space, which emphasizes the crucial importance of climate-related science. As such, NASA does not anticipate our core programming to deviate from the aims of the GWS.

Historically, NASA has carried out successful projects with numerous countries included on the HPC list, and it expects continued collaboration in those areas. NASA has found our applied science applications are far more impactful when they are co-developed with end-user participation. Accordingly, lack of end-user engagement could potentially impact the regions NASA is able to operate in, though NASA does not foresee any specific issues at this time. Additional risks inherent to the ongoing global COVID-19 epidemic or other potential pandemics could also impact our operations, though NASA has demonstrated a strong ability to adapt to a remote work scenario.

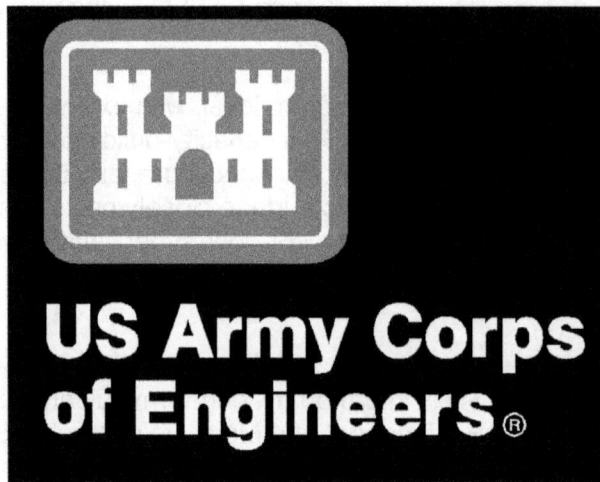

U.S. Army Corps of Engineers Plan

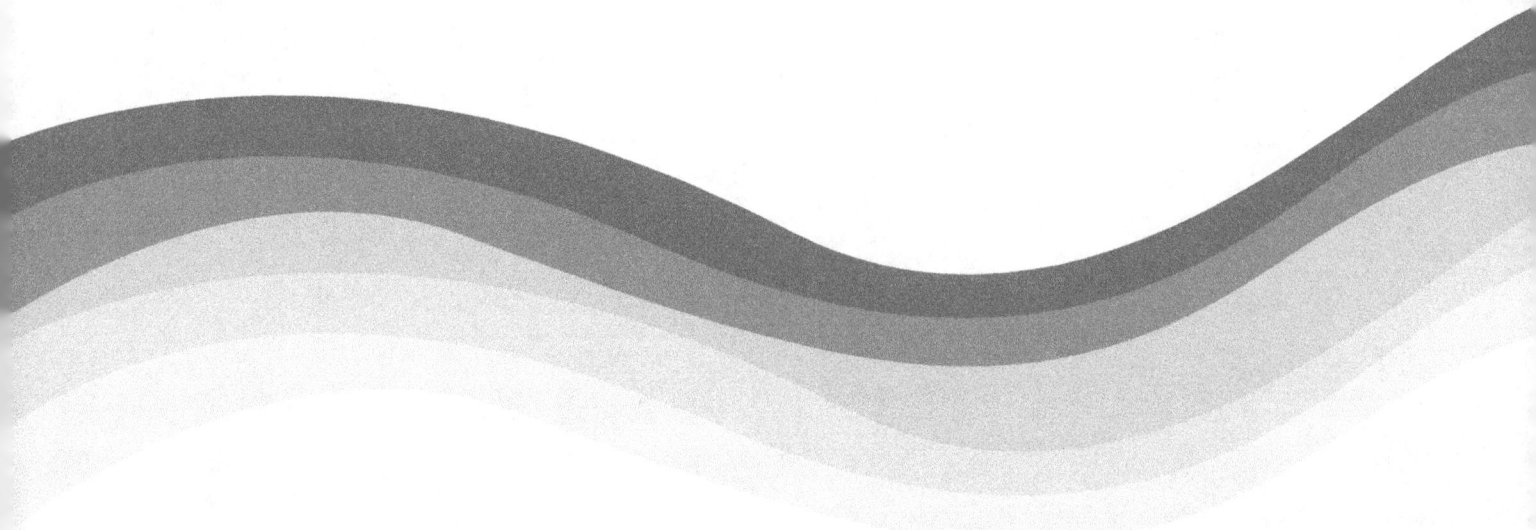

U.S. Army Corps of Engineers

Introduction

The mission of the U.S. Army Corps of Engineers (USACE) is to deliver vital public and military engineering services, partner in peace and war to strengthen our Nation's security, energize the economy, and reduce risks from disasters. USACE provides unique planning, technical, and managerial expertise to address domestic and international problems related to water resources, infrastructure development, emergency response, and environmental protection and restoration. USACE applies these capabilities in support of U.S. government departments and agencies, nongovernmental organizations (NGOs), international organizations, and foreign governments to address problems of national significance to the U.S.

USACE has authorities and skills to provide engineering and technical support to the Department of State, USAID, and other federal agencies whose international missions include aspects of water security. USACE also supports humanitarian assistance and disaster relief projects and activities on behalf of the U.S. Combatant Commands of the U.S. Department of Defense. This includes scalable integrated civil-military collaborative planning, innovative technologies, and integrated water resource management expertise to build resiliency and interoperability.

Another category of USACE support is to international organizations, foreign governments, and NGOs to address problems of national significance to the U.S. This support seeks areas of mutual benefit gained by partnering or working with non-U.S. government organizations to leverage the skills and resources of all parties. Under authority delegated to the Assistant Secretary of the Army – Civil Works, such activities are coordinated and must be consistent with U.S. Department of Defense goals and, if international, in consultation with the U.S. Department of State. This latter category of support is a special provision authorized through Section 234 of the Water Resources Development Act of 1996, as amended. USACE actively coordinates support across additional programs with matching strategic goals.

Contribution to the Global Water Strategy

The majority of USACE international water activities are driven by the U.S. Department of State and USAID requests for USACE core competencies. The U.S. Global Water Strategy (GWS) provides guidance that focuses USACE strengths in water resources management, including hydrologic extremes, climate change resiliency, risk assessment, and stakeholder engagement to serve the broader U.S. government interest.

Consistent with mission and authorities, USACE may offer contributions to the GWS SOs in the following areas:

Strategic Objective 1: Strengthen Water and Sanitation Sector Governance, Financing, Institutions, and Markets

Provide capacity building in the areas of water resources planning for national water ministries and local government agencies including increasing capabilities in disaster preparedness and response.

Strategic Objective 2: Increase Equitable Access to Safe, Sustainable, and Climate-Resilient Drinking Water and Sanitation Services, and Adoption of Key Hygiene Behaviors

Provide planning, design, construction management, and oversight of potable water and sanitation projects.

Strategic Objective 3: Improve Climate-Resilient Conservation and Management of Freshwater Resources and Associated Ecosystems

Provide capacity building in the areas of water resource management, including:

- Technical reviews, assistance, and advice on comprehensive water resources development;

- Assistance and training in water infrastructure risk identification, operations and maintenance; and operation of major river systems;

- Training in the use of up-to-date technical tools and skills to plan and optimally manage existing water resources assets and properly plan and invest in future water resources infrastructure;

- Comprehensive river basin and reservoir modeling and related assistance;

- Training and sharing best practices in dam safety management and supporting host nations with executing a comprehensive dam safety program inclusive of dam inspections and risk assessments and evaluating dam modifications;

- Predictive analysis and tools such as remote sensing for appropriate planning to respond to and mitigate impacts of a changing climate; and

Strategic Objective 4: Anticipate and Reduce Conflict and Fragility Related to Water

- Provide technical assistance in support of emergency responses initiated by a changing climate.
- Support public participation, collaboration, and conflict transformation initiatives related to water resource management, including marginalized and underserved stakeholder groups, through shared vision planning and related approaches.
- Provide technical assistance in the areas of transboundary water issues and conflict resolution.

Results Framework

USACE is committed to achieving results through our partnerships and individual reimbursable interagency agreements which are aligned with the SOs of the GWS. Similarly, USACE supports geographical U.S. Combatant Commanders and aligns with their unique lines of effort and focus areas. While USACE has no formal mechanism for tracking, aggregating, and reporting contributing GWS results, USACE does coordinate all results with each requesting agency consistent with the established goals in the applicable interagency and intra-agency agreement.

Approach

USACE is an executing agent, which means it conducts work at the direction of others and does not typically set policy or make unilateral decisions on the extent to which support activities will be employed. Other agencies leverage USACE capabilities, engineering solutions, and strong business practices to lead to desired outcomes such as self-reliance and sustainable improved conditions within a host nation. The most common approach is to use capacity building, which increasingly has gained worldwide recognition as being fundamental to effective governance, capacity enhancement, and enhanced ownership.

Principles

The operating principles as outlined in the GWS are supported through USACE's existing mission, vision, and priorities. USACE partners at all levels within a whole-of-government approach that is of mutual benefit. Through continued efforts to build capacity, USACE is often able to strengthen global, national, and local systems through strong partnerships, mentoring, training, education, and physical projects; the infusion of critical resources; and most importantly, the motivation and inspiration of people to improve their lives. Through seven premier research and development laboratories and multiple centers of expertise, USACE delivers innovative solutions to the nation's toughest challenges in military engineering, water resource management, civil works, geospatial research and engineering, and engineered resilient systems. This includes improving the safety and resiliency of communities and water resources infrastructure, and readiness to support disaster response, periods of instability, and emergency management planning and operations.

Geographic Focus

Geographical mission focus aligns with the U.S. Combatant Commands' geographical boundaries.

Resource Implications

Partner agencies leverage USACE capabilities on a reimbursable basis.

USDA

U.S. Department of Agriculture Plan

U.S. Department of Agriculture

Introduction

The U.S. Department of Agriculture (USDA) provides leadership, expertise, information, research, and programs to benefit U.S. agriculture, forestry, and other water users, and helps the U.S. supply high-quality products to the world. The Secretary of Agriculture provides leadership on agriculture, forestry, food, nutrition, natural resources, rural development, and related issues to advance sound public policy and efficient management, based on the best available science and evidence. Water-related expertise, programs, and tools exist throughout USDA because water security is fundamental to sustainable food, agriculture, forestry, and natural resources management systems. Water flowing from U.S. forests and grasslands provides important environmental, cultural, and recreational services that support rural livelihoods and resilience.

Coordination

The USDA Foreign Agricultural Service assists the Secretary of Agriculture on U.S. agricultural trade policy and international cooperation (7 USC §5693 and USDA Departmental Regulation 1051-002), coordinates USDA's participation in the Interagency Water Working Group, and facilitates the USDA Global Water Working Group to help USDA stakeholders coordinate their work to advance global water security.[16] USDA has offices at nearly 100 American Embassies covering approximately 180 countries. In countries where USDA has coverage, USDA Foreign Service Officers, Locally Employed Staff, or other USDA agency staff participate, as appropriate, in country-level teams that coordinate water issues and implementation of the U.S. Global Water Strategy (GWS).

Contribution to the Global Water Strategy

USDA contributes to domestic and international efforts to increase access to clean water and improve conservation, management, and restoration of watersheds and water resources through its international food assistance and capacity building programs; basic and applied research programs; environmental markets programs; data, analyses and information sharing; and the promotion of science-based policies and regulations that expand U.S. markets and trade. USDA has a long institutional history of collaborating with foreign governments, multilateral organizations, non-government partners (e.g., private sector, universities, and research institutions, international organizations), and other stakeholders as appropriate to achieve its mission. These USDA efforts will continue and, where practical, USDA's partnerships may be expanded and/or strengthened to help accomplish GWS objectives through the following framework of contributions and planned results.

---

[16] The USDA Global Water Working Group currently includes representatives from the Agricultural Research Service (ARS), U.S. Forest Service (USFS), Foreign Agricultural Service (FAS), National Institute of Food and Agriculture (NIFA), Natural Resources and Conservation Service (NRCS), Office of the Chief Economist (OCE), and Office of the Chief Scientist (OCS). These USDA offices and agencies also participate in the IWWG.

Framework

As summarized in the following table and explanatory notes, USDA's efforts concentrate on GWS Strategic Objective 3 (SO 3), though USDA contributes to all GWS SOs as appropriate.

- **Strategic Objective 1:** Strengthen Water and Sanitation Sector Governance, Financing, Institutions, and Markets
- **Strategic Objective 2**: Increase Equitable Access to Safe, Sustainable, and Climate-Resilient Water and Sanitation Services and the Adoption of Hygiene Behaviors
- **Strategic Objective 3:** Improve Climate-Resilient Conservation and Management of Freshwater Resources and Associated Ecosystems
- **Strategic Objective 4:** Anticipate and Reduce Conflict and Fragility Related to Water

USDA GWS Contributions	SO 1	SO 2	SO 3	SO 4
Sharing of relevant, publicly available information and analyses	Yes	Yes	Yes	Yes
Aligning of relevant international capacity building programs	No	Yes	Yes	No
Leveraging relevant domestic investments in water security	No	No	Yes	No
Providing technical and policy expertise on water	Yes	No	Yes	No

USDA will share its publicly available information on and analyses of global production, supplies, and demands of agricultural and forest commodities, and thereby continue to enable U.S. and foreign stakeholders to make better business and policy decisions for food security, nutrition, water, and other natural resources management. For example, USDA will continue to contribute to the Agricultural Market Information System to improve data collection and provide earlier warning of commodity price volatility that could affect water costs and vice versa. Through the National Agricultural Library and the Water and Agriculture Information Center, USDA electronically distributes scientific findings, educational methodologies, and public policy issues related to water and agriculture to scientists throughout the world. USDA also has decades of experience in international geospatial modeling initiatives, which help inform its Global Agricultural and Disaster Assessment System and USDA's World Production, Markets, and Trade reports and circulars, including monthly World Agricultural Supply and Demand Estimates. *Contributes to all SOs*

To the degree practical, USDA will align implementation of its international capacity-building programs with the GWS, including the Food for Progress Program, McGovern-Dole International Food for Education and Child Nutrition Program, Norman E. Borlaug International Agricultural Science and Technology Program fellowships, and Cochran Program fellowships. *Contributes to SO 2 and SO 3*

USDA will seek opportunities to leverage its existing relevant domestic investments in research, extension, education, and natural resources management through international cooperation. USDA will align its cooperative research activities with the GWS when applicable and appropriate. For example, the Soil and Water Assessment Tool (SWAT), jointly developed by USDA and Texas A&M AgriLife Research, is a physically based watershed and landscape model to simulate the quality and quantity of surface and groundwater and predict the environmental impact of land use, land management practices, and climate change. The SWAT is widely used domestically and internationally in assessing soil erosion prevention and control, non-point source pollution control, and regional management in watersheds. Another opportunity is the USDA Climate Hubs, which provide science-based, region-specific information and tools to agricultural and natural resource managers to make climate-informed decisions and build climate resilience. The Climate Hubs connect climate science to action through their operational framework and co-production process with stakeholders for effective development and delivery of climate information services. ***Contributes to SO 3***

USDA will support the GWS by making available, through consultations or other cooperative activities, its existing technical and policy expertise in sustainable agriculture, forestry, environmental markets, watershed and natural resources management, emergency management, climate change, nutrition, and rural development. USDA's expertise and programs can concurrently meet the needs of U.S. agriculture and forestry and, in some cases, support international efforts to improve water resources, availability, and management. In many cases, USDA provides expertise to State and USAID Bureaus via interagency agreements; for example, the U.S. Forest Service International Programs mobilizes experts to assist in watershed management, restoration and conservation, disaster preparedness and response, policy development, research, and trade issues with partners in over 90 countries. USDA experts also voluntarily participate in relevant Ambassador's Water Experts Program activities throughout the world. ***Contributes to SO 1 and SO 3***

Guiding Principles

USDA's approach and capacities to address international water needs reflect the guiding principles of the GWS. USDA's domestic focus to improve access to programs and services for underserved stakeholders and communities, as put forward in the Department's April 2022 Equity Action Plan, is also reflected in USDA's international work. As appropriate, USDA engages with partners at all levels of government and supports the development and dissemination of best practices. USDA's Department-wide 2022 Climate Strategy and Adaptation and Resilience Strategy, and USDA agencies' climate adaptation plans promote consideration and integration of climate change adaptation and resilience into USDA's work, including GWS efforts.

Geographic Focus

USDA's contributions to the GWS do not have a specific geographic focus.

Resource Implications

As appropriate and as indicated above, USDA will align its existing expertise, information, and programs with the GWS objectives. In addition, through interagency cost-reimbursable arrangements, USDA agencies could collaborate with USAID, the Department of State, or other U.S. government agencies to address GWS objectives.

Assumptions

As appropriate, USDA's implementation of the GWS will complement the GWS efforts of other U.S. government implementers as well as USDA's contributions to other water-related strategies and initiatives, including but not limited to the U.S. Global Food Security Strategy, the U.S. Action Plan on Global Water Security, the National Water Reuse Action Plan, the National Biodefense Strategy, the Agriculture Innovation Mission for Climate, the Plan to Conserve Global Forests: Critical Carbon Sinks, and the President's Emergency Plan for Adaptation and Resilience.

NOAA

NATIONAL OCEANIC AND ATMOSPHERIC ADMINISTRATION

U.S. DEPARTMENT OF COMMERCE

The National Oceanic and Atmospheric Administration Plan

U.S. Department of Commerce National Oceanic and Atmospheric Administration

Introduction

The U.S. Department of Commerce National Oceanic and Atmospheric Administration's (NOAA) three-part science, service, and stewardship mission is to understand and predict changes in climate, weather, oceans, and coasts; to share that knowledge and information with others; and to conserve and manage coastal and marine ecosystems and resources. NOAA has six programmatic areas, all of which contribute to water resources and water information in various ways. The National Satellite, Data, and Information Service (NESDIS) provides secure and timely access to global environmental data and information from satellites and other sources to promote and protect the Nation's security, environment, economy, and quality of life. The National Marine Fisheries Service (NMFS) provides stewardship of the Nation's ocean resources and their habitat. The National Ocean Service (NOS) provides science-based solutions through collaborative partnerships to address evolving economic, environmental, and social pressures on our ocean and coasts. The National Weather Service (NWS) provides weather, water, and climate data forecasts and warnings for the protection of life and property and enhancement of the national economy. The Office of Oceanic and Atmospheric Research (OAR) conducts research to understand and predict the Earth system, develop technology to improve NOAA science, service, and stewardship, and work with user communities to understand research needs and transition the results so they are useful to society. The Office of Marine and Aviation Operations delivers effective Earth observation capabilities, integrates emerging technologies, and provides a specialized, flexible, and reliable team responsive to NOAA and the Nation.

Water is a common thread that runs through NOAA's mission areas and offices, each of which serves stakeholders through a variety of field offices, laboratories, and national service outlets and engages internationally to further their missions, which aim to support healthy ecosystems, communities, and economies that are resilient in the face of change. With climate change acting as a central driver of ecological, social, and economic change worldwide, and water playing a pivotal role in the functioning of each, NOAA's work on water resilience is intimately linked with its work on climate change and climate resilience both domestically and abroad.

Contribution to the Global Water Strategy

Water is woven throughout NOAA's science, service, and stewardship mission. Science at NOAA is the systematic study of the structure and behavior of the ocean, atmosphere, and related ecosystems — a broad scope that includes the science of the global hydrologic cycle. In executing its Service mission, NOAA provides a range of Earth observations and scientific data sets, as well as information including water reports, forecasts, and warnings to inform the public, private, and academic sectors. Through its stewardship work, NOAA applies its knowledge to protect people and the environment, including through work to protect and restore freshwater habitats. Water also spans the full breadth of NOAA's research-to-application capabilities, including internationally, where its Line Offices contribute to water resources management to address issues including drought, extreme precipitation and flooding, water quality issues, and food insecurity. Through this Agency Plan, NOAA will bring the full breadth of the agency's water expertise – spanning global in-situ and satellite observations, fundamental research, weather and climate forecasting, bilateral and multilateral engagement, water resources management capacity building, support of humanitarian efforts around water-related food insecurity, and transboundary water cooperation – to bear in addressing water security issues worldwide.

NOAA's contribution to the Global Water Strategy is informed by several related strategies, policies, and plans. This Agency Plan is aligned with the 2022 White House Action Plan on Global Water Security, and outlines NOAA's contributions to implementing the Action Plan call for a whole-of-government approach to promote the sustainable management and protection of water resources and associated ecosystems. This Agency Plan is being developed in tandem with NOAA's Strategic Plan, which includes objectives related to water resources and prediction as well as NOAA engagement with international organizations on water issues. Additionally, NOAA's planned contributions to the Global Water Strategy are linked to the agency's planned contributions as one of the five core agencies responsible for implementing the President's Emergency Plan for Adaptation and Resilience (PREPARE), which aims to support developing countries and communities in vulnerable situations around the world in their efforts to adapt to and manage the impacts of climate change. NOAA's planned contributions involve climate information services, as well as climate adaptation work on infrastructure, water, health, and food security. Through commitments in these and other planning processes, water is among NOAA's highest priorities to contribute substantially to the advancement of society.

Results Framework

Strategic Objective 3: Improve Climate-Resilient Conservation and Management of Freshwater Resources and Associated Ecosystems

3.1 Observations, modeling, and decision-support

The NWS and the NESDIS carry out ongoing work together to provide leadership to the Group on Earth Observations Global Water Sustainability (GEOGLoWS) project. This product aims to achieve expanded operational capacity for water-related activities including observations, modeling, and decision-support applications through development projects, training, and partnerships among governments, academia, and participating non-governmental and international organizations directly supporting the water resources management objective of the Global Water Strategy.

NOAA supports the Integrated Drought Management Programme (IDMP), a global network co-sponsored by the Global Water Partnership and the UN World Meteorological Organization (WMO) designed to address issues related to drought monitoring, prediction, drought risk reduction, and management. IDMP helps to support stakeholders at all levels by providing policy and management guidance and by sharing scientific information, knowledge, and best practices for Integrated Drought Management. NOAA actively engages through this program to foster improved use of drought and climate science to improve early warning and proactive approaches to drought risk management.

OAR and NOS work in support of the Global Ocean Observing System (GOOS), a program under the Intergovernmental Oceanographic Commission of the United Nations (UN) Educational, Scientific and Cultural Organization that aims to achieve an integrated, responsive, and sustained global ocean observing system. While focused on ocean observations, GOOS plays a fundamental role in supporting improved weather and water variability predictions worldwide, since ocean conditions affect the terrestrial component of the hydrologic cycle. GOOS deliverables support end-users in better predicting weather and water variability as it affects agriculture, transportation, water management, energy production, and the lives of millions of people globally. For instance, through NOAA support, GOOS plays a critical role in developing requirements for sustained ocean and ocean-atmosphere observing, both of which have a direct impact on water budgets and their variability, and influence water resources management. NOAA works multilaterally with partners to leverage U.S. and international capabilities to support GOOS, for instance through work on the global Argo float and the tropical moored buoy system in the tropical Atlantic, Pacific, and Indian Oceans. In the Pacific, this observing system is essential to projecting the impact of the El Niño-Southern Oscillation (ENSO), a climate phenomenon that significantly influences weather and water variability. The GOOS 2030 Strategy seeks to deliver an integrated global ocean observations system that provides essential information for sustainable development, including water resources. To guide implementation of the GOOS Strategy, the global observing community developed SOs that outline key areas of activity to implement, including an effort to sustain, strengthen, and expand observing system implementation, promote standards and best practices, and develop metrics to achieve success by 2030. NOAA's international engagement in support of GOOS is an example of how the agency's provision of global ocean and atmospheric observations underpin the data and information necessary to understand and predict freshwater distribution changes that affect human welfare worldwide.

Link to action: NOAA's NWS, NMFS, and OAR along with U.S. government partner agencies will continue to work with the U.S. Department of State to internationally expand the implementation of Forecast-Informed Reservoir Operations (FIRO), a reservoir-operations strategy that better informs decisions to retain or release water by integrating additional flexibility in operational policies and rules with enhanced monitoring and improved weather and hydrological forecasts. FIRO intends to enhance water security and resilience in support of key Global Water Strategy principles.

3.2 Satellite programs and products

NESDIS runs the Joint Polar-orbiting Satellite System (JPSS), which provides global-scale observations of the hydrologic cycle that support water resources management worldwide. The JPSS program produces global water-related information products from the Suomi National Polar-orbiting Partnership (Suomi-NPP) and also creates products from a sensor on Japan's water observation satellite – Global Climate Observing Mission-Water 1st (GCOM-W1) – to observe critical variables in the global water cycle, including rain rate, snow cover and equivalent, sea-ice concentration, soil moisture, and precipitable water.

NESDIS also carries out ongoing work to run the CoastWatch/OceanWatch program, which monitors and distributes remote sensing satellite products, including for coasts and inland waters worldwide. This work supports SO 3 by improving water management outcomes through improved supporting datasets.

Link to action: NOAA Satellite data products are disseminated and used to inform disaster response worldwide – for instance, flood maps produced by JPSS are used by the International Disaster Charter, through which 17 space agencies and space system operators work together to provide satellite imagery for disaster monitoring purposes in more than 130 countries worldwide. An additional example of NOAA satellite applications for water resources management is the World Bank's South Asia Water Initiative, which uses data from NOAA satellites to inform its flood forecasting system for the Ganges and Brahmaputra Rivers and works to improve flood forecasting for the greater South Asia region.

Future implications of NOAA satellite programs: As extreme worldwide hydrological and meteorological events change in intensity, frequency, and distribution in the coming decades due to climate change, hydrological deliverables from NOAA's satellite programs will only gain in importance as the international need for improved forecasting and informed disaster monitoring grows. Pending continued baseline funding, NOAA will continue to deliver these critical water products internationally; with additional funding, the agency will be able to further innovate and refine its satellite offerings and international partnerships such as those outlined in this Agency Plan.

3.3 Fundamental research into the global hydrologic cycle

The Office of Oceanic and Atmospheric Research (OAR) carries out global hydrologic cycle research to achieve enhanced understanding of the mechanisms of drought, extreme precipitation, and water hazards, and improved early warning of water availability and water-related hazards. As an outcome of this work, NOAA provides data and services to international operational partners that help support risk-based decision-making for flood mitigation, drought resilience, water resources management, and emergency management.

Link to action: OAR will work closely with regional and national level institutions, including on water resources operations with the Caribbean Institute for Meteorology, and on improved resilience with the Caribbean Disaster Emergency Management agency. Partnerships will focus on the science, observations, and prototype applications needed to improve early warning for hurricanes, floods, droughts, and resilient infrastructure design for managing climate-related water risks, supporting SO 3 by improving resilience to climate-related hydrologic shocks and stressors. Pending additional funding, NOAA can continue to strengthen and expand its partnerships in the Caribbean focused on improving water resources operations. These services will be delivered by regional, national, and local entities working in partnership with NOAA. The success of these services will ultimately be measured by the increase in the number of applied products and partnerships, as well as the number of national and regional climate resilience plans and their implementation by partners in the region.

3.4 Weather and climate forecasting

The NWS runs its Global Forecast System (GFS), a weather prediction tool that couples atmosphere, land/soil, ocean, and sea ice models to predict weather conditions around the world. This effort supports SO 3 as NOAA provides weather and climate information internationally that is relevant to monitoring extreme conditions affecting water resources management.

Link to action: Currently, NOAA's GFS provides information for real-time, small-scale flash flood event guidance products to operational forecasters and disaster management agencies in 65 participating countries served by the WMO's Global Flash Flood Guidance System.

Implications of future innovation: NWS will implement its Next Generation Global Prediction System, which will increase the spatial and temporal resolution of predicted variables. NOAA improvements to this contribution will be measurable through improvements in both the spatial and temporal resolution of predicted variables shared with the international forecasting and disaster management community.

3.5 Bilateral and multilateral leadership and engagement

NOAA's Assistant Administrator for Weather Services is the Permanent Representative to the WMO and in this capacity coordinates U.S. engagement in the WMO's earth system approach to weather, water, and climate. The U.S. is participating in the WMO's Vision and Strategy for Hydrology and associated Action Plan for Water. This activity supports SO 3 since the WMO Action Plan for Water details activities to address floods, drought, hydrological data and science, water resources, and water quality. This work advances operational hydrology through enhanced science, infrastructure, capacity building, and related services, in the context of sustainable development and enhanced resilience. Ultimately, NOAA's success in engaging with the WMO on these topics can be measured in improvements in the capacity of National Meteorological and Hydrological Services (NMHS) internationally to deliver enhanced hydrological products and services.

Link to action: Through ongoing bilateral engagement with the Canadian Government, OAR and NOS provide leadership on the Great Lakes Coordinating Committee, which aims to leverage coordination between the U.S. and Canada to collect observations to sustainably manage the Great Lakes and St. Lawrence River for all users. This work is relevant to SO 3 as it results in hydraulic, hydrologic, and vertical control data that support water resources management. The long-term success of this work will be measured through the sustainability of the Great Lakes and St. Lawrence River for generations to come.

Link to action and future implications: NOAA and Brazil's National Center for Monitoring and Early Warning of Natural Disasters (CEMADEN) have collaboratively developed an MOU on Cooperation in Drought Monitoring and Early Warning, signed in May 2022. The MOU is a product of the U.S.-Brazil Joint Commission Meeting held in Brasilia in March 2020 and will remain in effect for ten years. The outcomes of this collaboration are expected to serve as guidance for the Brazilian federal government to improve emergency mitigation measures. CEMADEN and NOAA intend to bolster the existing framework for the drought monitoring and early warning system based on the U.S. National Integrated Drought Information System (NIDIS) and the CEMADEN monitoring and drought early warning system. Improved understanding of integrated observations in Brazil, including satellite-based monitoring for the prediction of drought onset and duration, will also improve drought monitoring in areas facing similar issues.

Strategic Objective 4: Anticipate and Reduce Conflict and Fragility Related to Water

4.1 Food security

In its ongoing work, NOAA cooperates with the U.S. Department of Agriculture (USDA), USAID, and others to support the Famine Early Warning Systems Network (FEWS NET) product, which aims to inform humanitarian assistance to drought-affected regions and supports SO 4 by providing support to reduce the human impacts of drought.

Link to action: In supporting the FEWS NET product, NOAA contributes the climate information on drought forecasts and severity that governments and relief agencies use in assessing risks to food security and planning for, and responding to, food-engendered humanitarian crises in countries in Africa, Central Asia, Central America, and the Caribbean.

4.2 Cooperation on transboundary waters

NOAA's NWS and OAR support the management of water resources that cross national borders, through the development of climate products and services as well as by providing help to determine data requirements and processing techniques and suitable presentation methods for predictions and products distributed to users of climatic, hydrologic, and meteorological information for transboundary waters.

Link to action: NOAA is involved in projects in the Colorado River (U.S. and Mexico) and Parana-Paraguay Basins (Bolivia, Brazil, Paraguay, Argentina, Uruguay) with the intent to improve weather, climate, and water forecasts (including runoff estimates and streamflow predictions) and promote cooperation on shared waters amongst transboundary stakeholders from these countries.

Principles

The Global Water Strategy principles are reflected in NOAA's approach, unique capacities, and contributions to implementing the Global Water Strategy. NOAA is focused on meeting the needs of marginalized and underserved people and communities, as outlined above in our contributions to the Group on Earth Observations Global Water Sustainability, the World Meteorological Organization, and the Famine Early Warning Systems Network. Together with these commitments, NOAA's water work in partnership with other agencies such as USGS, Bureau of Reclamation, USDA, USAID, and others, along with bilateral engagements with partner countries including with individual hydrometeorological agencies to improve local forecasting capabilities, showcases the agency's efforts to work through and strengthen global, national, and local systems.

As detailed in NOAA's contributions to improving climate-resilient water conservation and management and reducing water-related conflict and fragility, NOAA leverages research, learning, and innovation to serve as an authoritative source of weather, water, and climate data, information, and services internationally.

NOAA integrates resilience across its implementation efforts in this Strategy, whether that be through fundamental research into climate change effects on the global hydrologic cycle, drought modeling and projections, or collaborations with international partners to improve

6

disaster preparedness for the increasing frequency of extreme hydrologic events affecting vulnerable communities worldwide.

Geographic Focus

Much of NOAA's mission as it relates to water security spans the globe, serving countries facing water resource challenges such as food insecurity and flash flooding. For instance, the FEWS NET product that NOAA contributes to (section 4.1 for details) supports food security in countries in Africa, Central Asia, Central America, and the Caribbean, while NOAA's work as a project partner implementing the WMO Global Flash Flood Guidance System (section 3.4 for details) provides real-time guidance products to 65 participating countries in Africa, Asia, North, South and Central America, the Caribbean, the South-West Pacific, and Europe.

Resource Implications

NOAA's planned contributions to global water security, as described in this plan, represent core functions in the various Line Offices and therefore core budget items with significant allocations of agency funds and staff time.

Assumptions

The NOAA planned contributions outlined in this Agency Plan assume at least a flatline of funding to support existing Line Office programs with water deliverables, that full staffing requirements are met, and that no changes in the agency's strategic direction or factors that would limit the fulfillment of these contributions occur in the period 2022-2027. In addition, a reliance on additional future funding has been highlighted in the text where relevant.

U.S. Department of Defense Plan

U.S. Department of Defense

Introduction

The mission of the Department of Defense (DoD) is to deter war and ensure our nation's security. To do so, DoD maintains a Joint Force that sustains American influence and advances shared security and prosperity, together with our network of allies and partners.

The DoD is the largest U.S. agency and the nation's largest employer, with over 1.3 million active-duty service members, 750,000 civilian employees, and 811,000 National Guard and Reserve service members. In addition, more than 600,000 private sector employees are contracted to provide services and support to DoD, along with several hundred thousand workers in the defense industrial base. DoD military and civilian personnel operate in every time zone and climate, and more than 450,000 of our personnel serve overseas. DoD manages a global portfolio of over 568,000 buildings and structures, nearly 4,800 sites, and about 27.2 million acres of property.

Given its scale and footprint around the world, DoD recognizes the importance of global water security to ensure U.S. national security, provide for the safety and well-being of our personnel, and deter conflict. To execute our mission, DoD relies on military and civilian personnel and equipment being in the right place, at the right time, with the right capabilities, and this requires safe and dependable access to clean water across the globe. The availability, cost, and quality of water supplies are critical concerns for DoD infrastructure and operations, both within the U.S. and around the world.

To help address the linked threats of global climate change and water insecurity, DoD published the DoD Climate Risk Analysis (2021) and Climate Adaptation Plan (2021), which lay out the Department's strategic approach to addressing these interconnected challenges. Because the U.S. Global Water Strategy focuses on international water challenges, this plan does not consider water-related risks to DoD facilities and installations within the Homeland, but it does detail the contributions the Department will make to the U.S. whole-of-government effort to address water insecurity abroad within the scope of DoD's mandate.

Contribution to the Global Water Strategy

As highlighted in the White House Action Plan on Global Water Security, water security is essential to U.S. national security. Across much of the world, water security is under threat by a range of stressors, including accelerating climate change, creating new risks for international security. Water and climate security are key to DoD's ability to deter war and protect our country. The DoD Climate Adaptation Plan details how a failure to prepare for climate change, including implications for water resources, can erode military capabilities, degrade infrastructure, and result in missed opportunities for economic growth.

Water insecurity abroad, including the impacts of floods and droughts, can impede military operations, increase humanitarian crises, and create serious challenges for governments in providing essential social services and countering violent non-state groups. The DoD Climate Risk Analysis maps out the ways that water challenges abroad, along with the secondary and tertiary impacts of these challenges, can result in security concerns. For example, flooding can complicate access and basing, while drought can lead to competition over scarce water resources, heightening social and political tensions and contributing to instability, conflict, and/or the likelihood of mass migration. These water-related challenges may contribute to requests for U.S. government assistance, including from DoD, from partner countries, and/or directly threaten U.S. national interests.

Results Framework

This section of the Annex describes the ways in which DoD can contribute to the strategic objectives included in the U.S. Global Water Strategy. The U.S. Army Corps of Engineers prepared a separate U.S. Global Water Strategy (GWS) annex, and its civil works functions are not addressed in this annex.

Strategic Objective 1: Strengthen Water and Sanitation Sector Governance, Financing, Institutions, and Markets

This objective is outside the scope of DoD's mission.

Strategic Objective 2: Increase Equitable Access to Safe, Sustainable, and Climate-Resilient Drinking Water and Sanitation Services, and Adoption of Key Hygiene Behaviors

This objective is outside the scope of DoD's mission.

Strategic Objective 3: Improve Climate-Resilient Conservation and Management of Freshwater Resources and Associated Ecosystems

DoD is taking steps to improve the management of freshwater resources under a changing climate with international partner countries and on our installations around the world. Sharing water expertise between DoD and international partners can improve water resource management and contribute to water security, advancing the goals of the White House Action Plan on Global Water Security. DoD has worked with U.S. interagency partners to develop an operational Global-Hydrologic Information program, slated for roll-out in July 2023. This initiative will provide a readily accessible, authoritative source of global water information to assist in building allied and partner awareness, capacity, and resilience with which to enhance their water security.

Additionally, the Department's Climate Adaptation Plan details DoD efforts to identify the level of exposure of major installations globally to water hazards (e.g., floods and droughts), along with other climate hazards (e.g., rising sea levels). This analysis will inform energy, water, and climate-resilience planning, as well as emergency management plans. DoD is adopting climate-informed decision-making at all levels that makes use of analytic methods to estimate returns on relevant investments and the impact on the readiness of various climate adaptation and water-resilience measures.

Strategic Objective 4: Anticipate and Reduce Conflict and Fragility Related to Water

DoD will include water and climate security, and links to instability, fragility, and conflict, in relevant future risk assessments. In fiscal year 2022, DoD received appropriations for a global water security center (GWSC) that can support these efforts. The Department identified water security needs to guide GWSC research and analyses over the next several years on water and environmental security issues. This work will help achieve the goals of the White House Action Plan on Global Water Security.

Finally, DoD is revitalizing the Defense Environmental International Cooperation (DEIC) program to support international partners and allies in building environmental, water, and climate security resilience. In accord with the priorities and guidance of the Office of the Under Secretary of Defense for Policy, future DEIC-funded activities may include projects focused on water security, with an eye toward strengthening our strategic partnerships, improving allied and partner capabilities and capacity, and sustaining DoD mission resilience.

Geographic Focus

DoD's activities will focus on regions aligned with DoD strategic priorities as articulated in the National Defense Strategy (2022), the GWS, and the White House Action Plan on Global Water Security.

Resource Implications

DoD is working to request appropriate budget levels for DoD water and climate security efforts, consistent with the national security priorities laid out in the National Defense Strategy.

Environmental Protection Agency Plan

U.S. Environmental Protection Agency

Introduction

The mission of the U.S. Environmental Protection Agency (EPA) is to protect human health and the environment. EPA works to ensure that:

- Americans have clean air, land, and water;
- National efforts to reduce environmental risks are based on the best available scientific information; and
- Federal laws protecting human health and the environment are administered and enforced fairly, effectively, and as Congress intended.

To accomplish this mission, EPA develops and enforces regulations, gives grants, studies environmental issues, sponsors partnerships, teaches people about the environment, and publishes information.

EPA works to ensure drinking water is safe; oceans and watersheds are restored and maintained; and aquatic ecosystems protect human health, support economic and recreational activities, and provide healthy habitat for fish, plants, and wildlife through the implementation of statutes such as the Clean Water Act and the Safe Drinking Water Act.

EPA's water expertise is recognized globally. EPA is often approached by international water sector stakeholders – including government officials, the development community, and water providers – for help in the design and function of water systems abroad, including laws, science, and financial mechanisms. Contingent on resource availability, EPA shares U.S. domestic approaches, and elements thereof, internationally.

Contribution to the Global Water Strategy

EPA's Office of International and Tribal Affairs and Office of Water collaborate across EPA's other program offices and with other U.S. government (USG) agencies to develop and support international partnerships. EPA shares knowledge and expertise with U.S. and international water practitioners on joint areas of concern such as water quality management, laboratory quality assurance, water reuse, and water infrastructure finance.

Results Framework

EPA develops and implements programs or projects – for example, through interagency agreements or invitational travel funding – for activities that align with and complement the EPA strategic plan and are allowable international activities under U.S. laws and regulations. When EPA implements an interagency agreement or program, it puts in place specific and appropriate metrics based on the work and metrics developed by the agreement partner (e.g., the U.S. Department of State or U.S. Agency for International Development).

EPA approaches:

- Ensure project long-term viability consulting with stakeholders throughout the term of the project.

- Assure project results by putting in place regular communication, timelines, and budgets.

- Develop program management plans and logic models to explain program inputs, outputs and planned short-term, medium-term, and long-term outcomes.

Strategic Objective 1: Strengthen Water and Sanitation Sector Governance, Financing, Institutions, and Markets

- Develop sustainable and innovative water infrastructure financing mechanisms.

- EPA leads USG engagement on the Organization for Economic Development and Cooperation (OECD) Environment Policy Committee (EPOC), under which OECD's water finance activities fall. Working with OECD countries and sharing with the international water community of experts, EPA helps to guide these efforts in alignment with objectives and feedback from the greater USG federal family.

- Provide technical and policy guidance that are synergistic with domestic programs to ensure policy coherence between the U.S. and foreign programs. Working with the Department of State, U.S. Agency for International Development, Treasury, and others, advance U.S. interests with collateral environmental and foreign policy benefits by engaging in and tracking global water governance processes in the OECD, the World Health Organization (WHO), international finance institutions including multilateral development banks (MDBs) and Global Environment Facility (GEF), and the UN Environment Programme.

  - EPA supports WHO's guidelines on drinking water quality and guidelines for recreational water quality by evaluating chemical and microbial components, two critical considerations for ensuring safe water for human consumption and recreation.

  - EPA is the designated U.S. "Technical Focal Point" for the Land-Based Sources (LBS) Protocol under the Cartagena Convention and provides expert advice on management of LBS of pollution to both protect human health and coastal and marine resources.

  - EPA participates in UNEP's Global Partnership on Marine Litter by providing guidance based on EPA's domestic Trash Free Waters program. In 2016, EPA began a partnership with the UNEP Caribbean Environment Program to expand Trash-Free Waters to include international marine litter prevention initiatives, with Jamaica and Panama as pilot countries.

- Working with the USG and EPA priority countries and partners, EPA works to increase the capacity for environmental governance at the federal, state, or local level including toward the development of environmental laws and regulations, enforcement of environmental laws, and incorporation of public participation into programs.

Strategic Objective 2: Increase Equitable Access to Safe, Sustainable, and Climate-Resilient Water and Sanitation Services and the Adoption of Hygiene Behaviors

- On a reimbursable basis, support domestic and foreign training on water quality monitoring, laboratory capacity building, water safety plans to protect drinking water supply from source to tap, risk assessment and management of contaminants. Exchange information with governments and the international community on water quality assessments, monitoring, and development of health criteria, early warning systems, and international standards/guidelines such as the WHO drinking water, wastewater reuse, sanitation safety plans (SSP), and emerging contaminants (*i.e.*, cyanobacteria) to achieve health and safety targets.

- EPA is exploring and/or participating in several water, sanitation, and hygiene (WASH) partnership efforts, including:

  - Collaboration with International Finance Institutions, including MDBs and GEF, to improve water infrastructure finance mechanisms in emerging economies and to promote the water-related environmental performance of MDB efforts.

  - Country collaborations, including Brazil (water finance), Israel (water reuse), Ghana (water laboratory technical assistance), and the Philippines (water finance), which could be adapted and used as approaches in other places.

Strategic Objective 3: Improve Climate-Resilient Conservation and Management of Freshwater Resources and Associated Ecosystems

- EPA works with international partners to increase resilience including through the Educational Partnerships for Innovation and Communities (EPIC-N) and tools including the Climate Change Adaptation Resource Center (ARC-X).

- EPA provides best practices and lessons learned for transboundary water management and supports activities under the Great Lakes Water Quality Agreement and other U.S.-Canada and U.S.-Mexico transboundary issues in the restoration and protection of shared waters.

- EPA works through partnership- or place-based programs across the country to provide funding, technical assistance, and training to a variety of stakeholders within a watershed. These efforts are helping to provide the resources and support needed to address cross-cutting issues related to a changing climate (e.g., harmful algal blooms, ocean acidification, habitat conservation, etc.) and encourage collaborative water management solutions across jurisdictions and disciplines. Lessons learned could be shared internationally.

○ EPA's place-based, National Estuary Program (NEP) includes 28 estuaries managed by a variety of institutions including state and local agencies, universities, and individual nonprofits. In its oversight, EPA provides annual funding, national guidance, and technical assistance to the local NEPs. The 28 NEPs develop and implement Comprehensive Conservation and Management Plans shaped by priorities of local, city, state, federal, private, and non-profit stakeholders.

Approach

EPA's approach includes stakeholder consultation, development of program prioritization and logical frameworks with non-governmental and international organizations, municipal, state, and indigenous interests, and USG agency partners.

Principles

EPA's international priorities reflect how it supports and will apply the Global Water Strategy principles within its work. Namely, EPA strives to build the capacity of underserved communities and strengthen local approaches. EPA works to increase resilience and looks to leverage resources through these priorities:

- Combatting the climate crisis
- Advancing the values of environmental justice and equity
- Addressing transboundary pollution in North America and globally
- Building environmental infrastructure, a green economy, and green jobs and
- Strengthening environmental governance by deploying EPA expertise and innovation.

Geographic Focus

EPA works with countries, other USG agencies, and international organizations to determine geographic focus. Geographic focus depends on USG and EPA priority interests, ability, and interest of partners, or on a formal relationship such as a free-trade agreement.

Resource Implications

- EPA has robust water technical assistance capability and international experience applying those capabilities to support emerging economies, but EPA has limited internal resources to provide such assistance.
- EPA is open to exploring additional partnerships with international organizations, USG agencies, national and local governments, nongovernmental organizations (NGOs), community organizations, private sector entities, academia, and other relevant stakeholders to achieve shared international water objectives.
- Given EPA has limited resources, EPA is open to financial arrangements such as funding from other USG organizations or interagency efforts where the agency's knowledge can be adapted and transferred abroad, and for which resources are made available.

USGS

science for a changing world

U.S. Geological Survey Plan

U.S. Geological Survey

Introduction

The U.S. Geological Survey (USGS) mission is to monitor, analyze, and predict current and evolving dynamics of complex human and natural Earth-system interactions and to deliver actionable intelligence at scales and timeframes relevant to decision makers. As the Nation's largest water, earth, and biological science and civilian mapping agency, we collect, monitor, analyze, and provide science about natural resource conditions, issues, and problems. Our diverse expertise enables large-scale, integrated multidisciplinary investigations, and the provision of impartial scientific information to resource managers, planners, and our customers.

The USGS promotes the use of impartial scientific information by decision makers to: (1) minimize the loss of life and property as a result of water-related natural hazards such as floods, droughts, and land movement; (2) effectively manage groundwater and surface water resources for household, agricultural, commercial, industrial, recreational, and ecological uses; (3) protect and enhance water resources for human health, aquatic health, and environmental quality; and (4) contribute to wise physical and economic development water resources for the benefit of present and future generations.

Contribution to the Global Water Strategy

The USGS contributes to the U.S. Global Water Strategy (GWS) through activities conducted in collaboration with the Department of State, USAID, Department of Defense, and other Departments and Agencies that address international water science needs and improve the technical capacity of partner nations.

Water information is fundamental to national and local economic well-being, protection of life and property, and effective management of water resources. The USGS works with our partners to monitor, assess, conduct research, and deliver information on a wide range of water resources and conditions including streamflow, groundwater, water quality, and water use and availability. The USGS collects surface water, groundwater, water quality, and water-use data at approximately 1.9 million sites across all 50 states. These water data inform the publicly available National Water Information System and are accessible through the National Water Dashboard, Groundwater Watch, and additionally through cooperative platforms, including the Water Quality Portal and the National Groundwater Monitoring Network. In collaboration with the Department of State and USAID, the hydrologic methods, techniques, systems, software, and tools developed by the USGS to meet the needs of the Nation are either available or adaptable to address the water science needs of partner nations.

Results Framework

Strategic Objective 3: Improve Climate-Resilient Conservation and Management of Freshwater Resources and Associated Ecosystems

The USGS contributes to GWS SO 3 by leveraging domestic water science expertise, remote sensing and geospatial analysis capabilities, water data platforms, hydrologic models (e.g., MODFLOW), and other tools and methods with an emphasis on capacity building to increase the technical capabilities of partner nations. USGS capabilities and activities contributing to GWS SO 3 include:

1) Remote Sensing and field-based hydrologic, geologic, and geophysical data collection and mapping for assessment of water quantity and quality;

2) Assessments of water availability (quantity and quality), water use, and water-related hazards at local, regional, and national scales;

3) Hydrologic modeling to predict the consequences of water-related management actions (e.g., altered flow regimes caused by reservoir operations and diversions, groundwater withdrawals and coastal saltwater intrusion, exposure to agricultural chemicals, and naturally occurring contaminants);

4) Evaluation and analysis of surface water/groundwater interaction;

5) Operational support of the Landsat Missions and the Famine Early Warning Systems Network (FEWS NET);

   a) The joint NASA/U.S. Geological Survey Landsat series of Earth Observation satellites provide continuously acquired images of the Earth's land surface that assist managers and policymakers make informed decisions about natural resources and the environment;

   b) The USGS FEWS NET Data Portal provides access to geospatial data, satellite image products, derived data products, and software tools in support of FEWS NET drought monitoring efforts throughout the world;

6) Participation in the Group on Earth Observations Global Water Sustainability (GEOGLoWS) project and leadership in support of the GEOGLoWS vision of global water sustainability in support of the social, economic, and environmental health of nations;

7) Training on state-of-the-art methodologies for acquiring water resources information, including methods of data collection and analysis, quality assurance, and data management; and

8) Technology transfer, training, and institutional support of hydrologic system management for science-based decision-making.

Strategic Objective 4: Anticipate and Reduce Conflict and Fragility Related to Water

The USGS contributes to GWS SO 4 in support of Department of State and USAID efforts to address the transboundary water issues of international partners. The USGS provides remote sensing and hydrologic modeling expertise and extensive transboundary water science experience gained from participation in activities of the International Joint Commission (for waters shared by the U.S. and Canada), the Transboundary Aquifer Assessment Program, (evaluating priority aquifers along the U.S.-Mexico border), and support of the Mekong-U.S. Partnership *NexView* project working with USAID to promote good governance and transboundary cooperation in the Mekong River Basin. The availability of groundwater can sometimes reduce conflict, instability, and fragility. To that end, the USGS also collaborates with USAID, the Millennium Challenge Corporation, development banks, nongovernmental organizations (NGOs), and partner nations to assess groundwater availability in water-stressed nations while providing decision-support tools for effective management and use of these resources.

In addition, the USGS contributes to GWS SO 4 in support of Department of State and USAID efforts to address food insecurity. The USGS FEWS NET Data Portal aggregates open-source remote sensing data to deliver agri-climatology analyses for drought monitoring and crop forecasting purposes.

Principles

The USGS is committed to scientific integrity, the delivery of unbiased science, and the incorporation of best practices in sustainability and resilience, research, learning, and innovation. The USGS values the diversity of our workforce and colleagues and is committed to providing a safe, productive, and welcoming work environment at home or abroad. The USGS conducts international activities and scientific research in accordance with the Guiding Principles of the Global Water Strategy including working in partnership within a whole-of-government approach that strengthens global, national, and local systems and meets the needs of marginalized and underserved people, communities, and those in vulnerable situations. The USGS will collaborate with partner institutions to leverage USGS and cooperative research, innovation, and learning to the mutual benefit of the participants and will incorporate diversity, equity, inclusion, and accessibility (DEIA) in international operations and activities.

Geographic Focus

The USGS operates globally within the framework of international treaties, hazard-response agreements, and cooperative agreements and in support of needs identified by the Department of State, USAID, other federal agencies, partner nations, and NGOs. Although the USGS Landsat program operates globally and USGS FEWS NET activities are distributed regionally across drought-stricken areas, the USGS supports the prioritization of activities in the High-Priority Countries identified in USAID's Agency Plan.

Resource Implications

USGS international water science activities are primarily implemented through reimbursable funding agreements with the Department of State, USAID, other federal agencies, partner nations, and NGOs. The USGS will continue to explore ways to expand international activities and support the GWS in consultation with the Department of State, USAID, and USGS senior management.

Assumptions

- International needs align with USGS mission capabilities and priorities.
- USGS continues to receive funding from the Department of State, USAID, other federal agencies, partner nations, and NGOs at current or increased levels.
- Availability of the USGS staff and/or contractors required to develop and implement international agreements and projects.
- Travel of USGS scientists is not inhibited by ongoing or future pandemics, hazards, instability, or armed conflict.

U.S. International Development Finance Corporation Plan

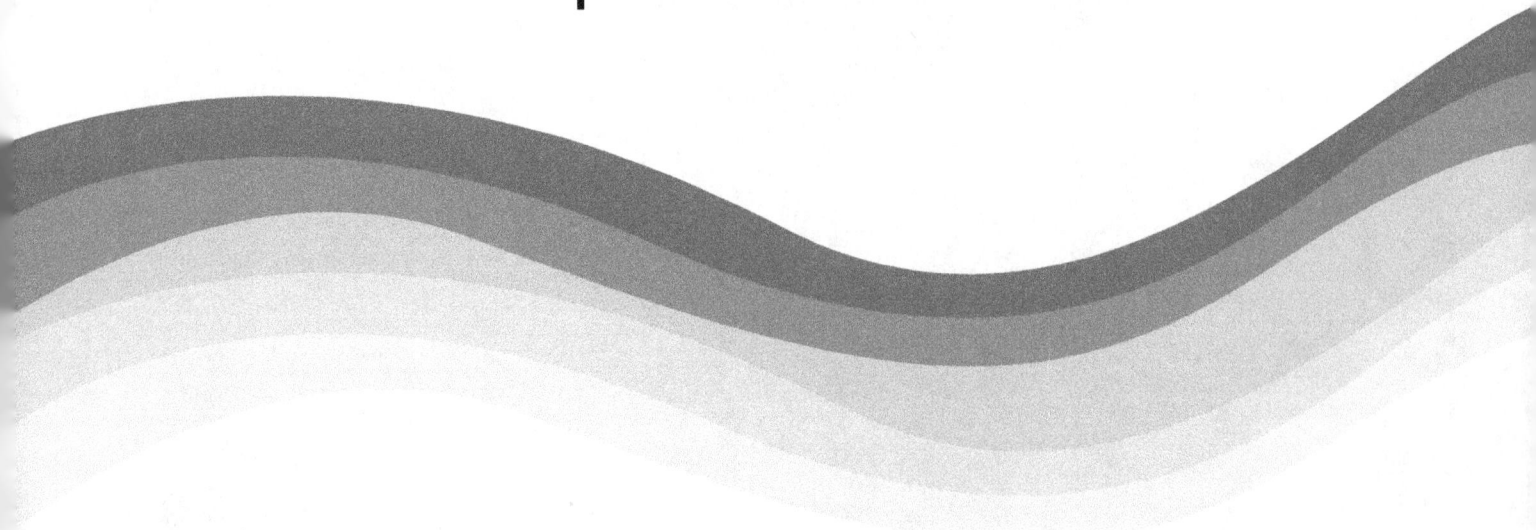

U.S. International Development Finance Corporation

Introduction

The Development Finance Corporation (DFC is the U.S international development finance institution. It focuses on supporting developmental, private sector investment in emerging markets. Its goal is to create critical development impacts in emerging markets and fragile states through provision of finance, political risk insurance, and technical assistance. Its financing tools include direct loans and loan guarantees as well as equity investments made through funds and directly to commercially viable, private sector businesses. Its political risk insurance includes inconvertibility of currency, expropriation, and political violence coverage as well as sovereign contract abrogation protection.

In support of U.S. foreign policy, economic, and strategic objectives, DFC prioritizes sectors including climate, gender equity, digital technology, and health. Financing and support for infrastructure, including water and sanitation projects, are important aspects of DFC's climate and health focuses. Water is a priority sector for DFC, and it has established goals for this sector. DFC's development strategy lays out the agency's goals for priority sectors, including a $500 million goal for water and sanitation projects by 2025. DFC aims to bring clean water to one million people by the end of 2025 and expects to establish similar goals for the ensuing years. The effects of DFC's investments that are expected to directly impact water access or water quality are determined annually and measured against the goals established. Projects in DFC's pipeline that are expected to have a positive impact on water access or water quality are tagged in DFC's databases.

DFC has a comprehensive water sector strategy that is intended to identify and pursue commercially viable private sector projects that meet its eligibility criteria. This plan has included researching key players in emerging markets, holding roundtables on investments in the water sector, conducting individual meetings with key stakeholders and investors, and following major water infrastructure projects globally to identify new opportunities. Financing opportunities that include debt and equity in water sector-focused investment funds are an important DFC focus. DFC's political risk insurance product has been attracting growing interest due to its ability to reduce perceived risks and enhance credit ratings through blue bonds and debt conversion for nature products. It was noted in DFC's recent Climate Event held on Earth Day 2022 that water projects are often key to climate adaptation and mitigation. Investments that contribute to meeting climate challenges will be of great interest to DFC in the future.

Contribution to the Global Water Strategy

DFC has a history of providing support for water sector projects including desalination plants, water pipelines, bottling facilities, agricultural projects with irrigation components, blue bonds as well as relevant health and sanitation projects. DFC currently has a strong pipeline of large and small private sector projects that include water treatment, irrigation, water resource management, sanitation systems, and marine and fisheries protection. Where water was primarily the responsibility of public sector entities in the past, there are new opportunities for private sector investment due to climate and health crises, innovative technologies, and a recognition that the private sector investment and innovative technologies will be critical to meeting global demand for clean water.

DFC expects to finance and provide political risk insurance for a wide variety of water and sanitation projects in the coming years. DFC lends to or provides equity directly or through investment funds in private sector projects that are expected to improve water access or water quality. DFC also offers political risk insurance that enables private sector investors to support water-related projects that may be exposed to risks of expropriation, currency inconvertibility, political violence, or sovereign government contract abrogation. DFC also provides technical assistance where applicable to ensure that projects will be commercially viable and sustainable while addressing water access and water quality issues. For some projects that are focused on water tangentially (manufacturing, agricultural processing, energy projects, health facilities, etc.), DFC may be able to provide technical assistance to improve their impact on water resources, access to clean water for employees and customers, or community access to new water resources.

DFC's debt, equity, and insurance products can be applied to projects alone, or in combination, where useful to enhance sustainability. These will range from wastewater treatment and reuse to drip irrigation systems and from water pipelines to bottling operations. It is also focused on providing technical assistance to projects to increase their ability to respond to climate-related impacts including loss of existing water resources; salinization of agricultural lands and aquifers; degradation of marine areas; inability to produce traditional agricultural products or raise animals due to extended droughts; and health crises that require increased hygiene, medical, and sanitation facilities.

Water is a crucial area of emphasis as it is integrated into many DFC sector strategies. DFC's Climate strategy includes a recognition of the relevance of the water sector to climate adaptation and resilience. DFC's health strategy includes an interest in sanitation and the clean water access needed to prevent and treat diseases. DFC's Food Security team is interested in the need for adequate irrigation for crops and clean water for animals. DFC's Technical Assistance team is expected to consider support to strengthen projects in terms of climate adaptation and resilience. Finally, DFC is also conscious of the impacts of water access loss or food insecurity on migration and civil disruption and will continue to support projects that address foreign policy, national security, and political and economic stability concerns.

Results Framework

Strategic Objective 1: Strengthen Water and Sanitation Sector Governance, Financing, Institutions, and Markets

In support of the U.S. Global Water Strategy (GWS), DFC will provide financing and political risk insurance in emerging markets for private sector projects that increase water access, water infrastructure, water and wastewater treatment and reuse, water conservation, and sanitation services. Through its debt and equity financing for the water sector, DFC strengthens the ability of emerging markets to meet the essential needs of their populations, increase economic growth, address climate impacts, and mitigate political instability. Examples of completed DFC-supported projects include a large desalination plant in Algeria, a water pipeline project in

Jordan, a water bottling project in the West Bank, a fund in Latin America that finances water infrastructure development in rural communities, and a debt conversion for nature project in Belize that will support marine conservation projects.

In addition to debt and equity financing, DFC has developed political risk insurance tools including debt-for-nature swaps and blue bonds that support projects involving the public sector and also address environmental issues related to safe and sustainable water resources. Blue bonds are financial instruments issued in the capital markets. Impact investors buy blue bonds with the knowledge that the funds raised will support investments that are related to expanding water access or supporting water resource conservation or restoration. DFC has a goal of committing $500 million in water sector financing by the end of 2025, and this DFC investment amount is expected to be increased significantly through private capital mobilization.

Strategic Objective 2: Increase Equitable Access to Safe, Sustainable, and Climate-Resilient Water and Sanitation Services and the Adoption of Hygiene Behaviors

DFC's financing and political risk insurance tools are applied to projects that address health, climate, the environment, and gender equity. In response to the loss of previously reliable water resources due to climate change and its impact on health and food security, DFC supports private sector investments in drip irrigation, water desalination, water and wastewater treatment infrastructure, and water pipelines. To address health needs amplified by COVID-19 and other communicable diseases, DFC supports investments in water-related equipment and services for health clinics, hospitals, businesses, and communities. Additionally, DFC addresses sustainable access through support for investments in water safety verification and efficiency measures including digital monitoring of water delivery systems, smart meters, water-efficient appliances, and HVAC systems in public and private buildings. DFC also anticipates support for investments in sanitation equipment and services that reduce illnesses and improve the health of disadvantaged populations, resulting in more equitable economic growth.

Strategic Objective 3: Improve Climate-Resilient Conservation and Management of Freshwater Resources and Associated Ecosystems

DFC expects to support climate-related investments in the management and protection of freshwater and marine resources, including nature-based solutions in critical environments. It has a record of supporting projects in conjunction with The Nature Conservancy and other entities through its debt conversion for nature projects and blue bonds. Projects that address the restoration of marine fisheries and damaged coastal environments are expected to remain important aspects of DFC's portfolio. With regard to improved management and protection of water resources, many DFC projects address climate adaptation and resilience by supporting projects that provide alternative water resources to farmers, commercial operations, and communities. As DFC focuses on increasing the amount of its investment that is relative to climate adaptation and mitigation in the next few years, water sector projects are expected to be a significant component as they often meet the criteria for climate adaptation and resilience.

Strategic Objective 4: Anticipate and Reduce Conflict and Fragility Related to Water

DFC invests in emerging markets and fragile states with the objectives of supporting economic development along with achieving U.S. foreign policy and national security goals. Where DFC creates alternative, clean water resources for communities that have lost access to water or where it has been an area of contention, DFC reduces the drive for migration and the root causes of some civil strife through its private sector-driven economic development. Where some political foes, terrorists, or other negative influences might use water conflicts to create political turbulence, the availability of reliable water resources can mitigate those disruptive efforts. DFC operates in both emerging markets and fragile states, with the latter being particularly subject to political and economic stability leading to migration impacts.

Principles

2022 Guiding Principles:

- Work through and strengthen global, national, and local systems
- Focus on meeting the needs of marginalized and underserved people and communities, and those in vulnerable situations
- Leverage data, research, learning, and innovation
- Incorporate resilience across all aspects of the strategy

The impacts of climate change, environmental fragility, political instability, shifting economic priorities, migration, and changing cultural practices can all impact water access, water quality, and water demand. DFC recognizes that the water sector will present both new challenges and opportunities in the coming years. DFC is taking a proactive approach to addressing these needs through its focus on climate adaptation and resilience, as well as direct investments in projects that build water pipelines and water treatment plants, improve access to water for agriculture and renewable energy, conserve freshwater and marine resources, and support water quality. In the coming years, DFC hopes to have individuals in place who can focus specifically on nature-based solutions globally, which should have a positive impact on water quality and water access. DFC will continue to invest in projects that directly impact areas where water is scarce and demand is rising, as well as providing increased technical assistance to projects in other sectors that could benefit from improved water usage practices to reduce water depletion in communities. DFC expects to be able to meet its goals for 2025 and accomplish similar goals in the ensuing years.

Focused on meeting the needs of populations in emerging markets, DFC prioritizes investments in low-income countries (LICs) and lower-middle-income countries (LMICs) where private sector investment may be most critical to ensuring good health, economic and political stability, and resilience. Wherever possible, the private sector projects that DFC supports include an effort to reach underserved populations. Investments directly or through financial intermediaries provide financing for rural communities and underserved urban areas to improve their water infrastructure. Where women spend large parts of their day burdened by seeking clean water for their families, ready access to community water facilities release them to focus on other productive activities.

Partnerships with agencies including State, USAID, and Millennium Challenge Corporation (MCC support holistic approaches to improving legal and regulatory environments. This is key to increasing private sector investments in fragile states and developing countries.

DFC's technical assistance program provides an opportunity to support investments with funding, information on best practices, and guidance on international standards. For many water sector projects, this technical assistance can strengthen their developmental impacts. It can also be applied to projects in non-water sector industries to support them in applying technologies that mitigate health or climate impacts related to water and sanitation.

The water sector provides some of the strongest opportunities for DFC to fund projects that are focused on climate adaptation and resilience. Where farmers have lost water resources or where over-salinization has depleted soils, DFC's support for water pipelines and innovative irrigation systems provide critical responses to climate impacts. Where communities have no access to clean water or sanitation systems, DFC investments in desalination, wastewater treatment facilities, or bottling operations can provide safe water and prevent disease outbreaks that limit people's ability to work, run businesses, or attend schools. Investments in digital technologies allow for water conservation and efficiency measures that help communities deal with newly limited water resources. From energy to ranching, from offices to homes, from hospitals to schools, water is an essential element of life. DFC's investments are supporting access to this life-saving resource.

Geographic Focus

DFC focuses its investments on the private sector in emerging markets and fragile states. Although it prioritizes investing in LIC and LMICs, DFC can also support highly-developmental projects in upper-middle-income countries. Projects in Latin America, the Middle East, Africa, and Asia will be targets of DFC support in the water sector. Water resources are often rated highly developmental because of the essential nature of water to health, climate adaptation, food security, economic growth, and political stability.

Resource Implications

DFC's Roadmap for Impact, its development strategy, lays out priority sectors that are considered to deliver the greatest developmental impact. Water is one of DFC's priority sectors. DFC is proactively pursuing projects in the water sector and has identified the top companies investing in the water sector in emerging markets for both direct and indirect outreach efforts. DFC staff support viable projects in the water sector using the agency's funds. There is no separate budget for water sector projects.

Goals established under the *Roadmap for Impact* set investments in the water sector at 500 million USD by 2025. It also has a goal of helping one million people to gain clean water access. Based on DFC's active portfolio and pipeline of potential projects, these appear to be achievable goals. With that in mind, DFC should meet and exceed that objective during the term of the Global Water Strategy which goes through 2028.

DFC is prioritizing climate-related projects. Water projects are often excellent examples of climate adaptation and resilience. As such, it is reasonable to expect that DFC will fully support a broad group of water projects in the coming years, as part of the GWS, with a significant portion of the overall climate goal being met through investments in the water sector.

DFC has a monitoring and evaluation process for each project that is managed by the Office of Development Policy. In addition, the Office of the Chief Development Officer tracks the progress toward goals laid out in the *Roadmap for Impact* for the priority sectors. In addition, assessments of the broad developmental effects created by the projects in the water sector are expected to be conducted.

Assumptions

DFC assumes that recent health crises, climate impacts, the introduction of innovative technologies, and the recognition by governments that they cannot meet the demand for new water resources will result in increased private investment in the water sector in emerging markets. DFC has a broader opportunity to support this investment as it is now able to provide equity and technical assistance for projects in addition to its traditional debt and political risk insurance tools. The fact that it can now finance projects that may not include a U.S. partner greatly increases the potential investments it can support. DFC assumes that it will have ongoing opportunities to work with impact funds, NGOs, International Finance Institutions, and other entities on climate, health, and gender-equity projects that will be important to expanding its portfolio in this sector.

DFC also assumes it may expand its reach in the water sector once it is able to provide local currency loans. Water projects are normally not hard-currency earners, so opportunities to borrow on a local currency basis should increase investment potential in many markets.

DFC assumes that it will be able to increase water sector projects based on its recent success in providing debt conversion for nature projects and blue bonds, as well as using debt and equity to support investment funds focused on the water sector.

The risks to expanding DFC's work in this sector are similar to those in other sectors. Increasing violence in fragile states or serious disruptions in key areas of emerging markets would limit private sector investment. DFC generally supports projects where there is private sector involvement and cannot dictate where those investors go. Any long-term health crises are also likely to cause serious economic impacts that could affect the commercial viability of water sector projects. Water projects that are focused on bringing water from distant areas to underserved populations could be disrupted by new climate impacts serious enough to limit water supplies at the intended source. Political unrest could also impact water resource access. Government retrenchment on privatizations could limit opportunities for private investment in key areas. The inability to offer local currency loans will have an inhibiting effect on some projects that DFC would hope to support.

USTDA
U.S. TRADE AND DEVELOPMENT AGENCY

Trade and Development Agency Plan

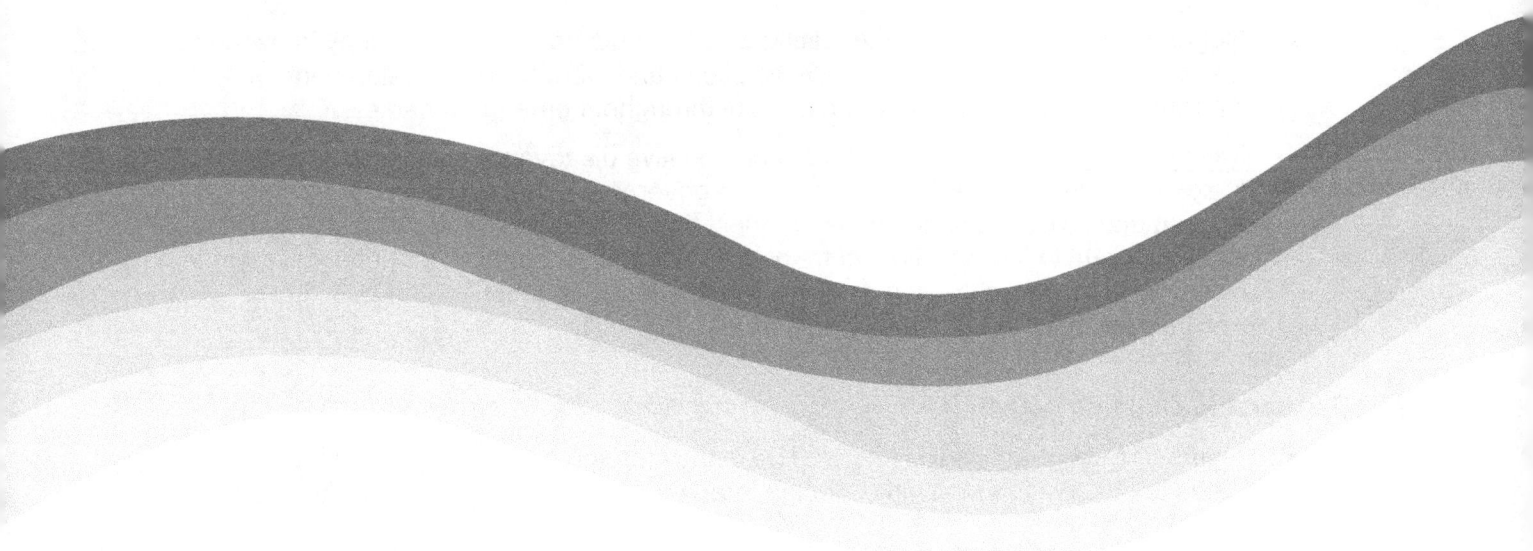

U.S. Trade and Development Agency

Introduction

The mission of the U.S. Trade and Development Agency (USTDA) is to help companies create U.S. jobs through the export of U.S. goods and services for priority infrastructure projects in developing and middle-income countries. USTDA links U.S. businesses to export opportunities by funding project preparation and partnership-building activities that develop sustainable infrastructure and foster economic growth in partner countries using the tools detailed below. For its partnership-building activities, USTDA conducts extensive market research and leverages its relationships to target these events to potentially imminent procurements or investments in infrastructure.

Contribution to the Global Water Strategy

Given USTDA's dual mandate to address priority infrastructure projects in developing and middle-income countries while also promoting the export of U.S. technologies and services, the Agency is particularly focused on supporting energy efficiency and information technology applications in service of the water sector. Since the Agency supports early project preparation activities, it also seeks to support activities that have a high probability of being financed and implemented by multilateral development banks, regional development banks, public and private sector financiers, and the U.S. government's (USG) own International Development Finance Corporation and Export-Import Bank.

USTDA's programmatic tools are contributions to Strategic Objective (SO) 1: Strengthen Water and Sanitation Sector Governance, Financing, Institutions, and Markets. Please see below a detailed description of each.

Grants

- <u>Project preparation assistance</u>: USTDA funds feasibility studies, technical assistance, and the comprehensive analyses that infrastructure projects need to move from concept to financing and implementation.

- <u>Pilot projects</u>: USTDA funds the testing of U.S. equipment and technology in overseas settings to promote cutting-edge U.S. solutions and identify new development opportunities for scalability and replicability throughout emerging markets.

- <u>Training grants</u>: U.S. firms may face a competitive disadvantage in tenders where a foreign competitor receives support from its government, helping them to exceed tender requirements. U.S. companies may request that USTDA help level the playing field with a training grant to fund the cost of training for the project sponsor that enables them to experience innovative training activities as an additional benefit of the U.S. firm's procurement offer.

Partnership-Building Activities

- <u>Reverse Trade Missions (RTM)</u>: USTDA brings foreign project sponsors to the United States to observe the innovative design, manufacturing, and operation of American products and services to support their infrastructure development goals. These strategically planned missions also present excellent opportunities for U.S. businesses to establish or enhance relationships with prospective overseas customers.

- Conferences and workshops: Conferences and workshops connect U.S. firms with foreign project sponsors. These sector or region-specific events are designed to showcase U.S. goods, services, technologies, and standards to foreign buyers and procurement officials. U.S. firms may also meet one-on-one with overseas project sponsors.

Initiatives

- Global Procurement Initiative (GPI): The GPI educates public officials in emerging markets on how to establish procurement practices and policies that integrate life-cycle cost analysis and best value determination in a fair, transparent manner. The GPI helps partner countries acquire high-quality, long-lasting technologies, while building smart, sustainable infrastructure with overall savings to their government. These procurement methods also open markets to greater international competition. Through GPI, USTDA deploys tailored support to partner countries. GPI experts have led trainings on green and sustainable procurement, clean energy technology, and achieving social and economic development outcomes through value-based procurement.
- Making Global Local (MGL): The MGL initiative streamlines USTDA's ability to share and support new global opportunities at the U.S. state and local level, giving the program a two-way capability to reach an alliance of partners. MGL recently added 15 new partners, now totaling more than 90 partners across 33 states who participate in the full range of USTDA's export-promotion programs.

Results Framework

USTDA's programming is anticipated to support SO 1 by strengthening water and sanitation sector markets and project financing opportunities in the sector.

USTDA medium-term activities that are anticipated to support SO 1 include:

Brazil Energy Efficiency and Smart Water Utility Studies

USTDA is funding two ongoing studies for COMPESA, the Brazilian state water utility in Pernambuco. The first is a feasibility study examining energy efficiency measures that could be implemented throughout the utility's water and sewage network. The second is a smart water utility technical assistance that will provide COMPESA with a 10-year roadmap of ICT solutions to improve its operational efficiency and quality of service.

Brazil Water Reuse Feasibility Study

USTDA is funding an ongoing water reuse feasibility study for SANEPAR, the Brazilian state water utility in Paraná. The study is supporting the development of a water reuse facility and construction of water pipelines that would supply recycled water to industrial customers.

Regional Latin America and the Caribbean Virtual Water Workshops

USTDA is funding a series of four virtual workshops focused on U.S. technologies and best practices that support priority water and wastewater infrastructure development projects in Brazil, Ecuador, Jamaica, and Mexico. Each workshop is tailored to the needs of the specific market and includes training by U.S. subject matter experts, water utilities, technology, and service solution providers, lenders, and procurement experts.

Uzbekistan Digital Twin Water Pilot Project

USTDA is funding a pilot project to provide support to Uzbekistan's national water company, UzSuvTaminot, in deploying a GIS-based digital twin model to track water usage and identify leaks in real-time to increase the efficiency of their water distribution system. The pilot project would involve a GIS system, asset management system, public engagement portal, and the installation of remote monitoring Internet of Things sensors.

U.S. GLOBAL WATER STRATEGY 2022-2027
ANNEX B: High-Priority Countries

ANNEX B: HIGH-PRIORITY COUNTRIES

Pursuant to Section 136 of the Foreign Assistance Act of 1961 (FAA), as amended by the Water for the World Act of 2014 (the Act), the U.S. government (USG) must focus its investments under the Act on those countries and geographic areas where the needs are greatest and to maximize impact and sustainability. To do so, the Act requires the USAID Administrator, via delegated authority from the President, annually designate no fewer than 10 water and sanitation High-Priority Countries (HPCs) based on specific criteria, in the context of U.S. foreign policy interests.

The following countries were designated HPCs for October 1, 2022 – September 30, 2023:

Democratic Republic of Congo	Mali
Ethiopia	Mozambique
Ghana	Nepal
Guatemala	Nigeria
Haiti	Philippines
India	Rwanda
Indonesia	Senegal
Kenya	South Sudan
Liberia	Tanzania
Madagascar	Uganda
Malawi	Zambia

The USG will implement this strategy in HPCs through country-specific plans, which will become available as digital appendices to this strategy and on GlobalWaters.org.

Designation of High-Priority Countries

To meet the requirements of the Water for the World Act, HPCs are designated through a rigorous four-step analytical process. The steps include:

1. Calculation of a data-driven WASH Needs Index score;
2. Calculation of a data-driven Opportunity Score;
3. Compilation of programmatic considerations; and
4. Identification of HPCs.

Per the Water for the World Act, HPCs will be the primary recipients of USG official water and sanitation development assistance.

Calculation of a Data-Driven WASH Needs Index Score

USAID develops a WASH Needs Index (Needs Index) score for all Independent States in the World as identified by the U.S Department of State Bureau of Intelligence and Research. In addition, USAID also completes a Needs Index analysis for any additional geographies in which

USAID works that are not recognized by the USG as independent states. The goal of the Needs Index score is to enable comparison across countries, using the methodology described below.

For each of these countries and geographies analyzed, USAID accesses the latest available data for the eight criteria identified in section 136(f)(1)(A)-(H) of the FAA, with some minor revisions needed to ensure alignment with the Sustainable Development Goals (SDGs). Additional details on these metrics, data sources, and notes are captured in the table below.

Once data have been compiled, USAID "normalizes" the score for each metric, or gives each score a comparable value between 0 and 1. To do this, USAID assigns the lowest score for each metric a value of zero and assigns the highest score for that metric a value of 1. All other scores are given a value between 0 and 1, depending on how close they are to the maximum or minimum of the range.

The overall Needs Index score is calculated by averaging the index scores for all eight metrics for each country. Each of the eight metrics is weighted equally, which results in 37.5 percent of the Needs Index focused on water access (three metrics), 37.5 percent of the Needs Index focused on sanitation access (three metrics), and the remaining 25 percent focused on under-five mortality due to diarrhea (two metrics). If data for any of the eight metrics are missing, no Needs Index score is calculated; this impacts a very small number of countries overall.

After the scores are calculated, the countries are ranked according to their overall Needs Index scores, with 1 being the country with the greatest degree of need. These rankings are used to identify the top 50 countries of greatest need as well as the top two quintiles (top 40 percent) of need to determine eligibility for the HPC designation process. The full needs ranking and underlying data are available in an interactive map.

Calculation of a Data-Driven Opportunity Score

In addition to considering needs, the Water for the World Act requires prioritization of those countries where water, sanitation, and hygiene programs have the opportunity to achieve maximum impact and long-term sustainability. Specifically, the Act requires consideration of:

1. Host-government commitment, capacity, and ability to work with the U.S. to improve water, sanitation, and hygiene, with a focus on building indigenous capacity as well as prioritizing and resourcing water, sanitation, and hygiene;

2. Opportunities to leverage existing public, private, or other donor investments;

3. The likelihood of making significant improvements in the health and educational opportunities available to women and girls as a result of access to WASH; and

4. Other criteria related to the furtherance of the Global Water Strategy.

A range of data sources and third-party indices are used to calculate the opportunity score for all countries included in the WASH Needs Index evaluation. Current data sources and third-party indices include the WHO/UNICEF Joint Monitoring Programme (JMP), WHO UN-Water Global Analysis and Assessment of Sanitation and Drinking-Water, the World Bank Country Policy and Institutional Assessment, the World Economic Forum Global Gender Gap data, and many others.

Following collection and analysis of the data, the aggregate scores for the four criteria noted above are equally weighted and averaged to determine the overall opportunity score.

Compilation of Programmatic Considerations

For all countries that are eligible for consideration as HPCs, country experts contribute to a summary of programmatic considerations that may not have been captured in the quantitative analysis process and impact the ability to achieve the objectives of the Water for the World Act. These considerations could include political and economic context and stability, U.S. mission capacity, and other considerations that could impact the ability of WASH programming to achieve maximum impact and long-term sustainability.

Identification of High-Priority Countries

The findings from the WASH Needs Index analysis, Opportunity Score, and Programmatic Considerations are further considered to identify HPCs. USAID requires that countries eligible for designation as an HPC:

1. ***Have "High" or "Medium High" needs according to the Needs Index***. This corresponds to the top two quintiles of Needs Index scores and ensures that investments focus on countries where the U.S. can help "provide sustainable access to clean water and sanitation for the world's most vulnerable populations," in alignment with the Water for the World Act; and

2. ***Be host to a bilateral USAID mission***. This ensures programs can "achieve maximum impact" as prioritized in the Water for the World Act.

At the conclusion of this analytical process, the USAID Water Leadership Council (see USAID Plan - Role and Responsibilities, Annex A) recommends countries for designation by the USAID Administrator via delegated authority from the President.

Needs Index Metrics, Data Sources, and Notes

Relevant Subsection of Water for the World Act	Metric	Data Source	Notes
136(f)(1)(A)	Proportion of the population without access to a basic drinking water service	JMP	This metric has been updated from proportion with an unimproved water source to proportion without access to a basic drinking water service to reflect the shift to the SDGs.
136(f)(1)(B)	Total population without access to a basic drinking water service	JMP	This metric has been updated from number with an unimproved water source to proportion without access to a basic drinking water service to reflect the shift in the SDGs.

Relevant Subsection of Water for the World Act	Metric	Data Source	Notes
136(f)(1)(C)	Proportion of the population without safely managed water access	JMP	This metric has been updated from piped water access to a safely managed drinking water service to reflect the shift in the Sustainable Development Goals. When sufficient data are not available, an estimate is made using available data on components of safely managed services (i.e., on-premises, available when needed, free from fecal and priority chemical contamination).
136(f)(1)(D)	Proportion of the population using shared or other unimproved sanitation facilities	JMP	N/A
136(f)(1)(E)	Total population using shared or other unimproved sanitation facilities	JMP	N/A
136(f)(1)(F)	Proportion of the population practicing open defecation	JMP	N/A
136(f)(1)(G)	Total number of children younger than five years of age who died from diarrheal disease	WHO Global Health Observatory	For countries without data available, data is pulled from the Institute for Health Metrics and Evaluation (IHME)
136(f)(1)(H)	Proportion of all deaths of children younger than five years of age resulting from diarrheal disease	WHO Global Health Observatory	For countries without data available, data is pulled from IHME

Notes

[i] United States National Intelligence Council, "Intelligence Community Assessment: Global Water Security." February 2012.

[ii] OECD, "States of Fragility 2020."

[iii] United Nations Environment Programme (UNEP), Food and Agriculture Organization of the United Nations (FAO), and Oregon State University "Atlas of International Freshwater Agreements." 2002.

[iv] Strong, C., S. Kuzma, S. Vionne, and P. Reig. "Achieving Abundance: Understanding the Cost of a Sustainable Water Future". 2020.

[v] Prüss-Ustün, A., J. Bartram, T. Clasen, J. Colford, O. Cumming, others. "Burden of Diarrheal Disease from Inadequate Water, Sanitation and Hygiene in Low- and Middle-Income Countries: A Retrospective Analysis of Data from 145 Countries." *Tropical Medicine and International Health* 19, no. 8 (2014): 894–905; Freeman, M. J. Garn, G. Sclar, S. Boisson, K. Medlicott, K. Alexander, G. Penakalapati, D. Anderson, A. Mahtani, J. Grimes, E. Rehfuess, T. Clasen. "The impact of sanitation on infectious disease and nutritional status: A systematic review and meta-analysis." *International Journal of Hygiene and Environmental Health* 220, no. 6 (2017): 928–949.

[vi] Global Commission on Adaptation and World Resources Institute, "Adapt Now: A Global Call for Leadership on Climate Resilience." 2019.

[vii] IPCC, "Intergovernmental Panel on Climate Change report". 2021.

[viii] Larsen, D., T. Grisham, E. Slawsky, and L. Narine. "An individual-level meta-analysis assessing the impact of community-level sanitation access on child stunting, anemia, and diarrhea: Evidence from DHS and MICS surveys." *Neglected Tropical Diseases* (June 8, 2017); Jung, Y., W. Lou, and Y. Cheng. "Exposure–response relationship of neighborhood sanitation and children's diarrhoea," *Tropical Medicine and International Health* (April 27, 2017).

[ix] UNICEF/World Health Organization (WHO) Joint Monitoring Programme for Water Supply, Sanitation and Hygiene (JMP), "Progress on Drinking Water, Sanitation and Hygiene: 2000–2020, Five Years into The SDGs." 2021.

[x] UNICEF/WHO JMP, "Progress on Drinking Water, Sanitation and Hygiene in Schools, 2000-2021 Data Update." 2022.

[xi] UNICEF/WHO JMP, "Fundamentals First: Global Progress Report on WASH in Health Care Facilities." 2020.

[xii] IPCC, "Summary for Policymakers." In: *Climate Change 2022: Impacts, Adaptation, and Vulnerability. Contribution of Working Group II to the Sixth Assessment Report of the Intergovernmental Panel on Climate Change.. Cambridge University Press. 2022. In Press.*

[xiii] Damania, R., S. Desbureaux, A.S. Rodella, J. Russ, and E. Zaveri. "Quality Unknown: The Invisible Water Crisis.", 2019.

[xiv] UNICEF. "Fast facts: WASH in conflict." 2019; UNICEF. "Water Under Fire Volume 1: Emergencies, development and peace in fragile and conflict-affected contexts." 2019.

[xv] Adelphi, International Alert, The Wilson Center, Institute for Security Studies, "A New Climate for Peace: Taking Action on Climate and Fragility Risks." 2020.

[xvi] USAID, "Suggested Approaches for Integrating Inclusive Development across the Program Cycle and in Mission Operations: Additional Help for ADS 201." 2018.

[xvii] Strong, "Achieving Abundance."

[xviii] Sanitation and Water for All. "Water and Sanitation: How to Make Public Investment Work: A Handbook for Finance Ministers." 2020.

[xix] Kolker, J., B. Kingdom, S. Trémolet, J. Winpenny, and R. Cardone "Financing Options for the 2030 Water Agenda. Water Global Practice Knowledge Brief." 2016.

[xx] Wolf, J. et al. "Impact of drinking water, sanitation and handwashing with soap on childhood diarrhoeal disease: updated meta-analysis and meta-regression." (March 14, 2018); Jung, "Exposure-response relationship"; Larsen, "An individual-level meta-analysis."

[xxi] WHO and UNICEF. "State of the World's Sanitation: An urgent call to transform sanitation for better health, environments, economies and societies." 2020.

[xxii] Hutton, G. L. Haller, and J. Bartram. "Global cost-benefit analysis of water supply and sanitation interventions." *Journal of Water and Health* 5, no. 4 (2007): 481-502.

[xxiii] Pickering, A.J. and J. Davis. "Freshwater availability and water fetching distance affect child health in sub-Saharan Africa." *Environmental Science & Technology* 46, no. 4 (2012): 2391–2397; Venkataramanan, V., J.L. Geere, B. Thomae, J. Stoler, P.R. Hunter, S.L. Young "In pursuit of 'safe' water: the burden of personal injury from water fetching in 21 low-income and middle-income countries." *BMJ Global Health* 5.10e003328 (2020).

[xxiv] Wolfe, J., S. Hubbard, M. Brauer, A. Ambelu, B. Arnold, R. Bain, V. Bauza, J. Brown, B. Caruso, T. Clasen, J.M. Colford Jr, M. Freeman, B. Gordon, R. Johnston, A. Mertens, A. Prüss-Ustün, I. Ross, J. Stanaway, J. Zhao, O. Cumming, S. Boisson. "Effectiveness of interventions to improve drinking water, sanitation, and handwashing with soap on risk of diarrhoeal disease in children in low-income and middle-income settings: a systematic review and meta-analysis" *The Lancet* 400, no. 10345 (July 2–8, 2022): 48-59.

[xxv] Graham, J., M. Hirai, and S. Kim. "An analysis of water collection labor among women and children in 24 sub-Saharan African countries." PloS ONE 11, no. 6: e0155981 (2016); Sorenson, S., C. Morssink, P. Campos. "Safe access to safe water in low-income countries: water fetching in current times." *Social Science & Medicine* 72, no. 9 (2011): 1522–1526.

[xxvi] Peters, K. and L. Peters. "Disaster Risk Reduction and violent conflict in Africa and Arab states Implications for the Sendai Framework priorities." 2018; UN Water & UNESCO, "The United Nations World Water Development Report 2019: Leaving no one Behind." 2019.

[xxvii] IUCN, "Fact Sheet: Disaster and gender statistics." n.d.

[xxviii] Nagabhatla, N., T. Avellan, P. Pouramin, M. Qadir, P. Mehta, J. Payne et al. UNESCO world water assessment programme (WWAP). "The United Nations World Water Development Report 2019: Leaving No One Behind." Paris: UNESCO, 2019: 44–57.

[xxix] Ide, T. M. Ensor, V. Le Masson, S. Kozak. "Gender in the Climate-Conflict Nexus: Forgotten" Variables, Alternative Securities, and Hidden Power Dimensions." *Politics and Governance* 9, no. 4 (2021).

[xxx] "Cholera—*Vibrio cholerae* infection." CDC website. Accessed March 22, 2022; World Meteorological Organization. "Atlas of Mortality and Economic Losses from Weather, Climate and Water Extremes (1970–2019)." 2021.

[xxxi] Faour, G. and A. Fayad. "Water Environment in the Coastal Basins of Syria—Assessing the Impacts of the War." *Environmental Processes* 1, no. 4 (2014); Gleick P. *The World's Water* 8, (2014): 148; Schillinger, J. G. Özerol, Ş. Güven-Griemert, M. Heldeweg. "Water in war: Understanding the impacts of armed conflict on water resources and their management." (2020).

[xxxii] United States Government. "Economics of Resilience to Drought in Ethiopia, Kenya and Somalia." Washington, D.C.: United States Agency for International Development Center for Resilience, 2018.

[xxxiii] "Global WASH Cluster - Humanitarian Response Dashboard 2021." WASH Cluster website.

[xxxiv] OECD. "States of Fragility Framework." 2018.

[xxxv] UNICEF. "Water Under Fire."

[xxxvi] OCHA. "Global Humanitarian Overview 2022." 2021.

[xxxvii] USAID. "Bureau for Humanitarian Assistance Emergency Assistance Guidelines—Annex A." 2022.

[xxxviii] Zeitoun, M. and N. Mirumachi. "Transboundary water interaction I: reconsidering conflict and cooperation." *International Environmental Agreements: Politics, Law and Economics* 8, no. 4 (2008).

[xxxix] Wolf, A. "Shared waters: Conflict and cooperation." *Annual Review of Environment and Resources.* 32 (2007): 241–269.

[xl] Bukari, K., P. Sow, and J. Scheffran. "Cooperation and Co-Existence Between Farmers and Herders in the Midst of Violent Farmer-Herder Conflicts in Ghana." *African Studies Review* 61, no. 2 (2018).

[xli] USAID. "Building Resilience into a River Basin." 2021.

[xlii] USAID. "Programming Considerations for HDP Coherence." 2022.

[xliii] USAID. "Humanitarian-Development Coherence in WASH or WRM Programs." 2021.

[xliv] Levy K., A. Woster, R. Goldstein, E. Carlton. Untangling the Impacts of Climate Change on Waterborne Diseases: a Systematic Review of Relationships between Diarrheal Diseases and Temperature, Rainfall, Flooding, and Drought. *Environmental Science & Technology.* (2016) May 17;50(10):4905-22.

[xlv] Ryan S.J., C.A. Lippi, F. Zermoglio. Shifting transmission risk for malaria in Africa with climate change: a framework for planning and intervention. *Malaria Journal.* 2020 May 1;19(1):170.

[xlvi] Blum A.J., P.J. Hotez. Global "worming": Climate change and its projected general impact on human helminth infections. *PLoS Neglected Tropical Diseases* (2018) 12(7): e0006370.

[xlvii] MacFadden, D.R., McGough, S.F., Fisman, D. *et al.* Antibiotic resistance increases with local temperature. *Nature Climate Change* (2018) **8,** 510–514.

[xlviii] Burnham J.P. Climate change and antibiotic resistance: a deadly combination. *Therapeutic Advances in Infectious Disease.* (2021) Feb 15;8

[xlix] WWAP (UNESCO World Water Assessment Programme). The United Nations World Water Development Report 2019: Leaving No One Behind. 2019.

[l] UNESCO and UN-Water. United Nations World Water Development Report 2020: Water and Climate Change. 2020.